ALEX KIMMONS

THE HISTORY OF THE ATOM
THE PERIODIC TABLE
and
RADIOACTIVITY

PREFACE

Chemistry is an interesting and fundamental branch of science because it gives us the chance to explain the secrets of nature. What is water? What do we use in our cars as fuel? What is aspirin? What are perfumes made of? These kinds of questions and their answers are all part of the world of chemistry. There is no industry that does not depend upon chemical substances: the petroleum, pharmaceuticals, garment, aircraft, steel and electronics industries, for example, as well as agriculture, all utilize the science of chemistry. This book helps everyone to understand nature. However, one does not need to be a chemist or scientist to understand the simplicity within the complexity around us.

The aim was to write a modern, up-to-date book where students and teachers can get concise information about the structure of substances. Sometimes reactions are given in detailed form, but, in general, excessive detail has been omitted.

The book is designed to introduce fundamental knowledge in three areas: the history of the atom, the periodic table and radioactivity. We will study the historical development of atomic structure theories, the tendencies of elements in periods and groups, and the types of emissions and uses of radioactivity.

CONTENTS

THE HISTORY OF THE ATOM

INTRODUCTION

The things we see when we look around us - trees, animals, ships, televisions, computers, etc - all consist of tiny particles. These particles are so small that we can not see them with the naked eye.

Democritus (460–400 BC), a Greek philosopher, first proposed that the universe was made up of very small particles and named them 'atomos', which means 'indivisible' in Greek. Aristotle, however, thought that matter was continuous, that there was no limit on how finely you could cut it up.

The ratio between sun and a man is approximately equal to the ratio between a man and an atom by mass.

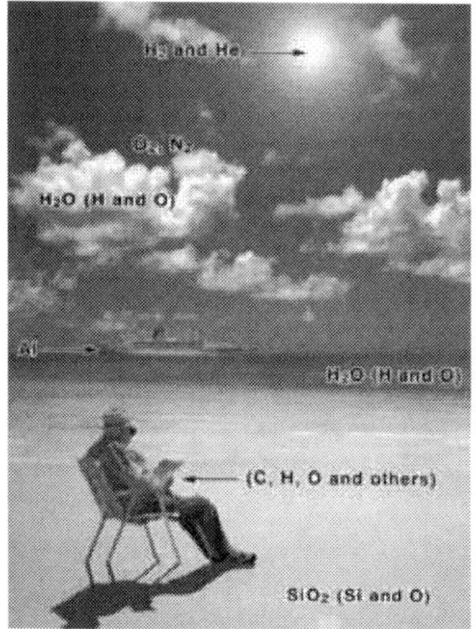

All substances consist of small particles called atoms.

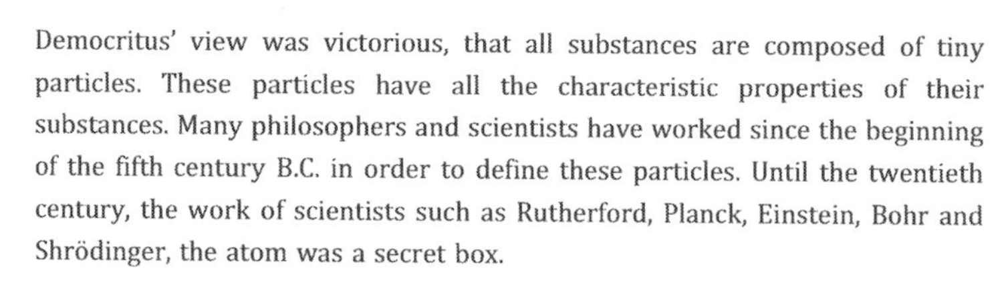

Democritus' view was victorious, that all substances are composed of tiny particles. These particles have all the characteristic properties of their substances. Many philosophers and scientists have worked since the beginning of the fifth century B.C. in order to define these particles. Until the twentieth century, the work of scientists such as Rutherford, Planck, Einstein, Bohr and Shrödinger, the atom was a secret box.

Nowadays, the presence of atoms can easily be shown with the help of today's technology, either by utilizing computer facilities which can visualize atoms or through their behaviors in different chemical reactions.

READING Seeing the Atom

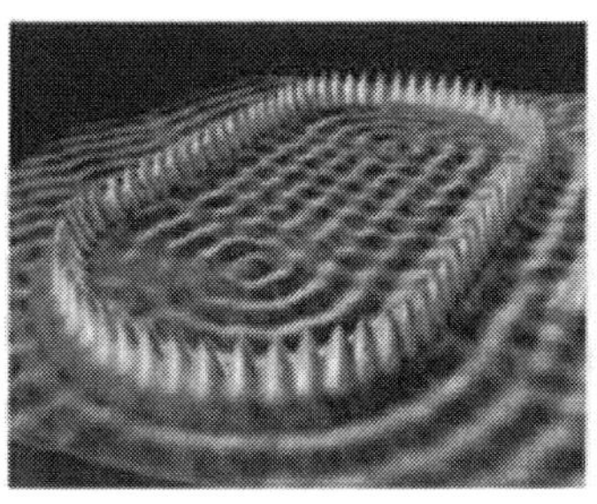

Iron atoms are placed on copper (Picture is enlarged about one billion times.)

When the first microscope was invented about 400 years ago, human beings were faced with an amazing world. New studies of the micro world followed. Single biological cells were observed in detail, and viruses were imaged by the invention of the electron microscope. With rapidly developing technology, we were introduced to new types of microscopes. One of them was the Scanning Tunnelling Microscope (STM) for which Gerd Binning and Heinrich Rohrer were awarded with the Nobel Physics Prize in 1986.

How does STM run?

STM is used to image the surfaces of substances with high electron conductance. The device is based on the system in which solid substances are covered with a microscopic atmosphere composed of electrons. The thin probe of STM is very sensitive. It collides its atmosphere with the surface of the electron atmosphere by getting very close, around 10^{-8} cm away. When this happens, a small electric current is created because of electron leakage between the surface and the probe. This current is proportional to the distance of the probe from the surface. Holding the current constant, the probe is moved right to left and up and down across the surface. As a result, projections of the dimensions of an atom and even cavities can be detected. The projections are fed into computers and three-dimensional, colored images of the surface obtained. These images are not real views of atoms, but they are the probable equivalent of a tunnelling map which consists of surface position versus height.

STM in industry:

STM has wide application areas in basic industry related to atomic surfaces. For example, they are used in producing high quality hard disc recorder headers. STM also provides the possibility of observing electrochemical reactions and atomic spectrum.

STM can be used in electrolytic solutions, liquid helium, oil, water and airless conditions. Because of these specifications, it can be also used to image DNA or working cell electrodes.

The Scanning Tunnelling Microscope

1. ATOMIC MODELS

What we know today about atoms is the product of long term studies in both theoretical and experimental concepts. Firstly, after the acceptance of the presence of atoms in order to search and investigate their properties, various atomic models were proposed.

Many atomic models were imagined by scientists, but these were not ideas proposed from the direct observation of atoms. One of the models was like grapes distributed throughout a cake, another like the solar system and another like the layers of an onion.

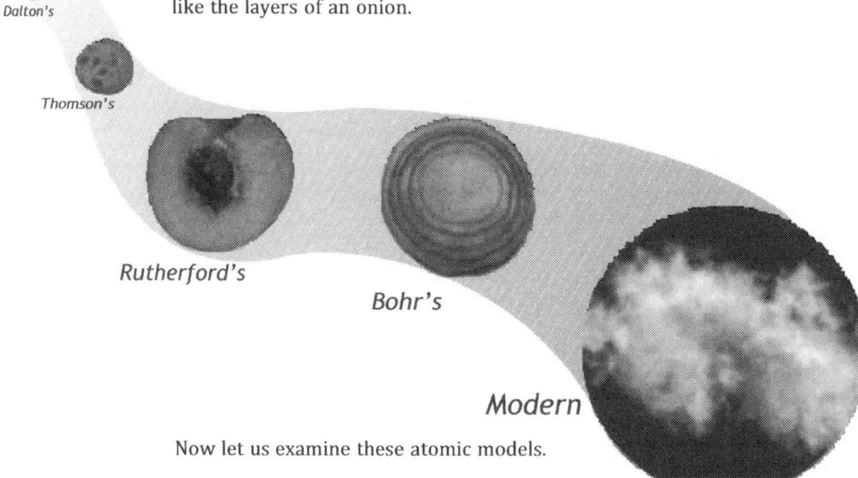

Now let us examine these atomic models.

1.1 DALTON'S ATOMIC MODEL

In 1803, by utilizing the mass relationships seen in the laws of either definite or multiple proportions, Dalton proposed that substances are a community of smaller particles. The fundamental properties of Dalton's theory were:

Dalton's proposal of atomic model

1. The unit particles showing all of the characteristic properties of substances are the atoms or groups of atoms.
2. Atoms of an element are completely identical.
3. An atom is a filled sphere, like a billiard ball.
4. Different types of atoms have different masses.
5. The smallest building bricks of substances are atoms. Atoms can not be further divided.
6. Atoms form molecules by combining in a definite numerical ratio. For example; one atom of substance X combines with one atom of substance Y to form the compound XY, while one atom of X combines with two atoms of Y to form the compound XY_2.

READING — The Cathode Ray Tube and Television

The color tube of the television that we use at our home is simply a vacuum tube, or cathode ray tube (CRT). CRT's are frequently used in televisions, computer monitors or anywhere where it is necessary to produce a picture.

The working mechanism of a CRT can be summarized as follows:

1. The Electron Gun

The electron gun is simply a hot metal (cathode) emitting electrons from the back of the tube. There are three electron guns in a CRT each of which corresponds to one of the three primary colors (red, green and blue). The electromagnetic signals coming from the transmitting TV antenna reach these electron guns at the back of the TV. The powerful guns then separate these signals into the three primary colors and send them to the fluorescent screen. The number of electrons and their intensities passing through the medium can be changed or regulated by applying a variable negative voltage to the grills found between the hot cathode and the anode; hence, the brightness of a point on the fluorescent screen can easily be changed.

2. Reflecting Planes

In one CRT, there are two types of reflecting planes.

a. *Vertical reflecting plane pairs: since the reflecting plane pairs are in a vertical position they reflect the incoming electrons horizontally.*
b. *Horizontal reflecting plane pairs: since the reflecting pairs are in a horizontal position, they reflect the incident beams of electrons vertically.*

   With the help of these two different reflecting planes it is possible to make horizontal and vertical electron sweeps on the fluorescent screen, and the images are thus created instantly. The fluorescent surface is scanned 16,000 times in one second.

3. Fluorescent Screen

The inner surface of the glass on the wide end of the CRT is coated with a fluorescent material. The surface of the screen is coated with a huge number (approximately 200,000) of phosphorus dots so that when an electron falls on the screen, these dots sparkle. The chemical composition of phophorus dots are oxides of rare earth metals.

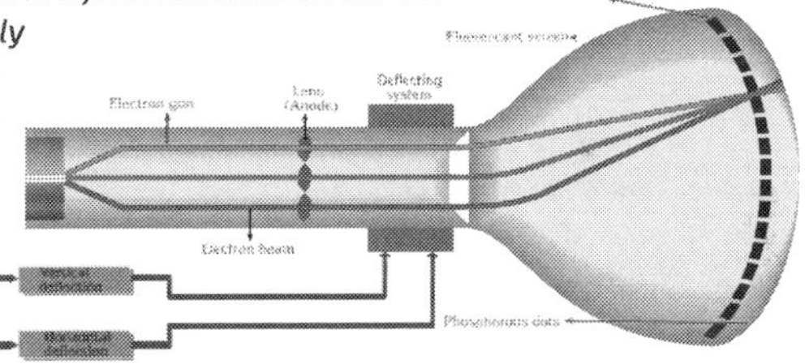

When a highly energetic electron coming from the electron gun strikes a phosphorus dot on the TV screen, the kinetic energy of the electrons turns into visible light in the form of a sparkling phosphorus dot. The electrons inside the phosphorus dot are excited by this strike. Then, when these already excited electrons fall back to their ground state, they give off their excess energies in the form of photons of a definite wavelength and frequency. That is why we see that point as a specific color. The reflecting and scanning systems mentioned above scan and show pictures dot by dot. The quality of the picture on the screen (resolution) depends on the speed of the scanning electrons and their accuracy in striking the phosphorus dots on the TV screen. Although a picture on the TV screen seems to be composed of a number of colors, actually all the colors are derived from the mixture of these three primary colors (red, green, blue). For instance, if we mix red with green we get yellow, and by mixing the three primary colors (red, green, blue), the color white is obtained.

1.2. THOMSON'S ATOMIC MODEL

Dalton's atomic model does not include negatively charged (–) electrons and positively charged (+) protons. Thomson discovered the electron in 1897, while the discovery of the proton was made by Rutherford in 1919. We can summarize Thomson's ideas as follow:

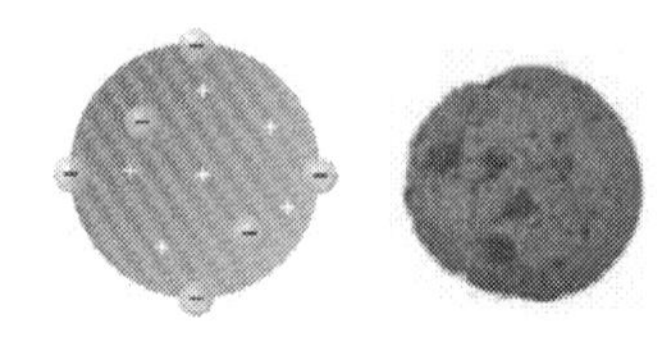

Thomson's proposed atomic model

1. Protons and electrons are charged particles. They have the same charge with opposite signs. The proton is equal to a charge of 1+ and the electron is equal to 1–.
2. In a neutral atom, since the number of protons and electrons are the same, the total charge is zero.
3. An atom has the shape of a sphere with 10^{-8} cm radius. In this sphere the protons and electrons are occupied in an arbitrary position. The distribution of electrons through this sphere resembles grapes distributed throughout a cake or plum pudding.
4. The mass of electrons is so small that it can be neglected. For this reason the mass of the atom is almost equal to the mass of the protons.

One of the defects of Thomson's atomic model is that neutrons were not mentioned in any way. The arbitrary distribution of protons and electrons throughout the atom was also not true.

1.3. RUTHERFORD'S ATOMIC MODEL

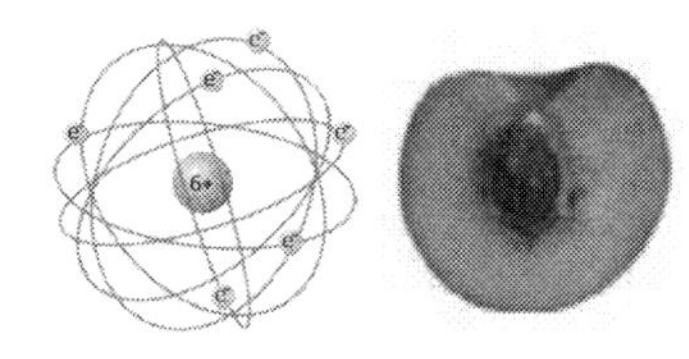

Rutherford's proposed atomic model was just like the solar system, with small bodies (the planets, etc) orbitting the sun due to the attraction force exerted on them by a large body (the sun).

Rutherford made important contributions to the explanation of atomic structure. He discovered the nucleus in 1911 and the proton in 1919. Prior to Rutherford, Thomson's atomic model was valid. His model stated that the atom was a sphere in which electrons and protons were moved arbitrarily. But there was an important question about how these protons and electrons were distributed. Was there any regularity or were they moving arbitrarily? The answer to this question could not yet be seen. In order to get answers to these problems and to verify Thomson's atomic model, Rutherford proposed a model resulting from his α – particle experiment.

In this experiment, Rutherford found that the vast majority, 99.9%, of the α – particles sent onto the metallic plate did not deviate or if they did, this deviation was very small so that they passed directly through the plate. He also noticed that a few of these particles were reflected back after striking the metallic plate. After that, Rutherford repeated his experiments using lead, copper, and platinum plates instead of gold, and obtained the same results.

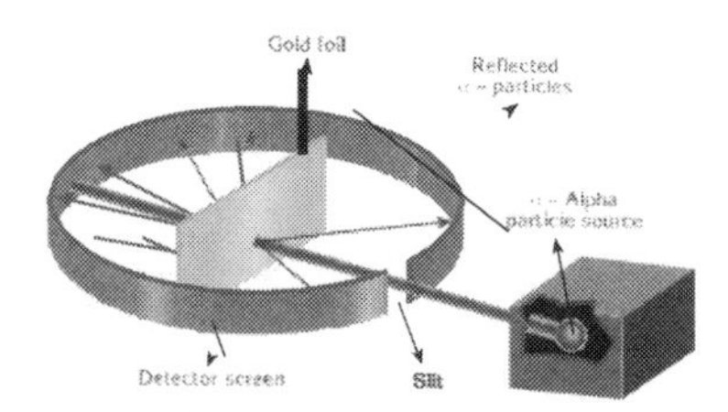

Rutherford's α - particle scattering experiment

The nucleus is very tiny compared with the rest of the atom. If the atom was the size of a football stadium, the nucleus (at the center) would be the size of a pea.
If the atom had no empty space, the size of the world would be the size of a football and the mass of the world would be unchanged.

From these results, Rutherford proposed the following statements:

1. There is a small, positively charged, dense region in the atom, a theoretical region which Rutherford named the 'nucleus'.
2. The mass of the atom is approximately equal to the mass of the protons and electrons.
3. According to Rutherford's atomic structure, the protons inside an atom are all gathered at the centre (nucleus) of the atom, with the electrons scattered randomly around, as the fruit of a peach is formed around the stone in the center, or the solar system, in which the sun is the nucleus and planets are electrons.

1.4. BOHR'S ATOMIC MODEL

Rutherford`s model stated that positively charged protons were found in the center of the atom, the nucleus . Negatively charged electrons were found around the nucleus. However, it did not indicate how electrons were arranged outside of the nucleus of an atom. In his theory, Bohr examined the movement of electrons.

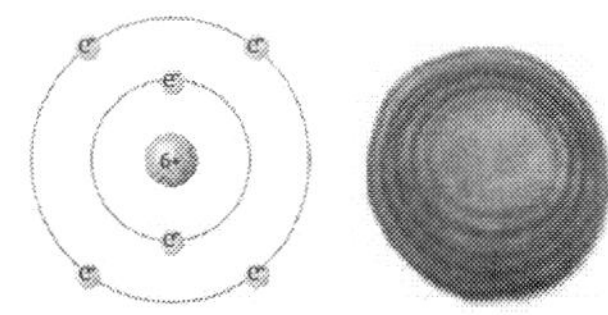

Bohr's proposed atomic model

In 1913 Niels Bohr proposed his atomic theory with the help of the line spectrum of hydrogen atoms and Planck's quantum theory. His postulates can be summarized as follows:

1. The electrons in an atom move at a certain distance from nucleus and their motions are stable. Each stationary state has a definite energy.
2. Electrons move in each stationary energy state in a circular path. These circular paths are called 'energy levels' or 'shells'.
3. When an electron is in a stationary state, the atom does not emit (radiate) light. However, when an electron falls back to a lower energy level from a higher one, it emits a quantum of light that is equal to the energy difference between these two energy levels.
4. The possible stationary energy levels of electrons are named either by letters, K, L, M, N, O ... or by positive integer numbers starting from the lowest energy level. These numbers are generally denoted by "n" where n = 1, 2, 3, ...

In the light of today's knowledge Bohr's concept of the movement of electrons in a circular path is wrong.

1.5. THE DISCOVERY OF THE NEUTRON

In the 1930's, the Rutherford proposals of subatomic particles (positive charges) forced to Chadwick to do some experiments. Chadwick thought that if the positive particles (protons) occupy the center of an atom, then since same charged particles would repel each other and the structure of atoms would eventually be destroyed. However, the atom existed and it had a structure. If that was so, Chadwick thought that "there must be some other particles holding these protons together in a very small volume".

He named these particles 'neutrons', which means that they have no charge. The mass of a neutron is about $1.6749 \cdot 10^{-27}$ kg. If we remember that the mass of the proton was $1.6726 \cdot 10^{-27}$ kg, we can see that the mass of the neutron is a little bit bigger than that of proton. In nature, the only element which has no neutrons is hydrogen, and its mass is therefore equal to the mass of its proton.

THE PIONEERS

James Chadwick(1900-1958)

Chadwick was a famous English physicist. In the year 1932, he proposed that in the nucleus of an atom there are also electrically neutral particles having equal mass as protons. He gave the name 'neutron' to these particles, meaning 'has no charge'. For this discovery he won the 1935 Nobel Physics Prize.

2. MODERN ATOMIC MODEL

Bohr proposed his theory by assuming that electrons are mobile, charged particles and that the electron in a hydrogen atom can only have some certain energy levels. Although this theory was valid for atoms like hydrogen with one electron, such as He^{1+} and Li^{2+}, it does not work for atoms having more than one electron. It did not explain their atomic spectra and chemical properties, although it was a step forward in the development of modern atomic theory.

Modern atomic theory depends on the application of developments in wave mechanics to the movement of electrons. The pioneers of this theory are Lois de Broglie, Heisenberg and Schrödinger.

In 1924, considering the nature of the light and matter, Lois de Broglie proposed that small particles sometimes show wave-like properties. In 1927, his hypothesis was proved by the deviation of electron beams like X-rays by a crystal.

About 1920, Werner Heisenberg examined the affects of light in order to determine the behaviors of particles smaller than the atom. As a result, Heisenberg proposed his uncertainty principles.

Even if the position of a particle is known, it is impossible to determine exactly where the particles come from and where they go.

In other words, the position and the speed of the electron can not be determined at the same time. Because of this, it cannot be said that electrons

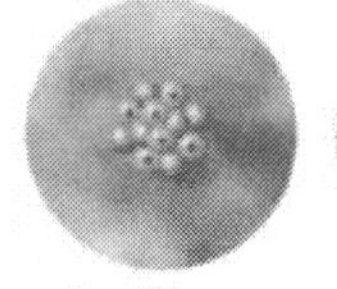

Modern proposed of atomic model

revolve around the nucleus in the certain circular paths (orbitals). Instead of orbitals, it is more accurate to talk about the probability of finding an electron (or electrons) around the nucleus.

Modern atomic theory explains the structures and behaviors of the atom much better than the earlier atomic theories. This theory explains the probability of finding electrons around the nucleus by virtue of quantum numbers and orbitals. The quantum numbers are the integer numbers designating the energy levels of the electrons in an atom, and the orbitals are the probable regions in which the electrons might be found around the nucleus.

When the electron is considered as a particle, the orbital designates the region in which the probability of finding electron is high. On the other hand, when the electron is considered as a wave, the orbital represents the region in which electron charge density is high. That is, when the electron is considered as a particle, the probability of finding an electron at a definite point is specified, whereas when the electron is considered as a wave, electron charge density is specified.

In the light of this information, we will start to discuss the four quantum numbers, the types of subshells and the symbols representing the subshells.

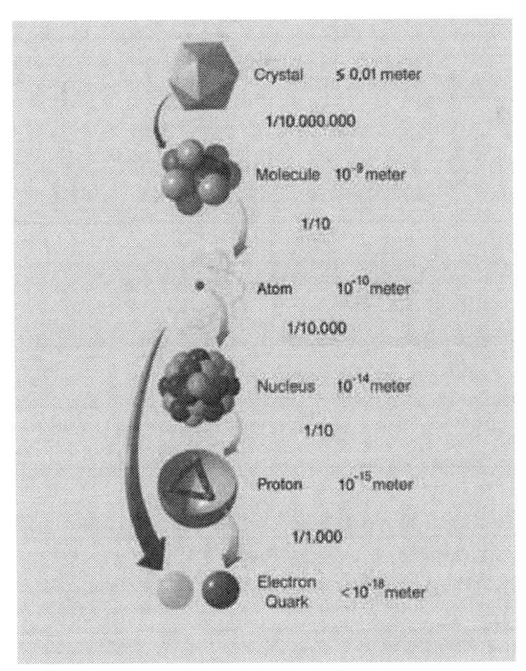

From matter to quarks: a schematic representation.

2.1. QUANTUM NUMBERS

Quantum numbers specify the address of each electron in an atom. There are four types of quantum numbers:

1. Principal quantum number, n → energy level (shell)
2. Secondary quantum number, l → subshell (s, p, d, f)
3. Magnetic quantum number, m_l → orbital
4. Spin quantum number, m_s → spin type of electron

There are no two electrons in an atom that can have the same four quantum numbers. Each electron has a unique address, like a family living in a flat. This is Pauli's Exclusion Principle.

Let us consider the address of a family. Imagine that a family migrates and seeks a new flat in another district. In that district there are K, L, M, N...streets which contain s, p, d, f... apartments with one, three, five and seven floors respectively and two flats in each floor. That family is not a rich one and there are many kinds of flats with varying rents. When a street or apartment is changed the rent to be paid is changed too. Flats in K street have the lowest rents, and then L, etc. The family could decide, for example, to live in ' L ' street, ' p ' apartment, p_y floor and 'clockwise ' flat.

The table below shows this approach:

Family:	Electron
District:	Quantum numbers
Street:	Principal quantum number (shell) (K, L, M, N , ...)
Apartment:	Subshell (s, p, d, f)
Floor:	Orbital (s, p_x ,p_y ,p_z ,...)
Flat:	Spin type (clockwise or counterclockwise)
Rent:	Energy level (in any address)

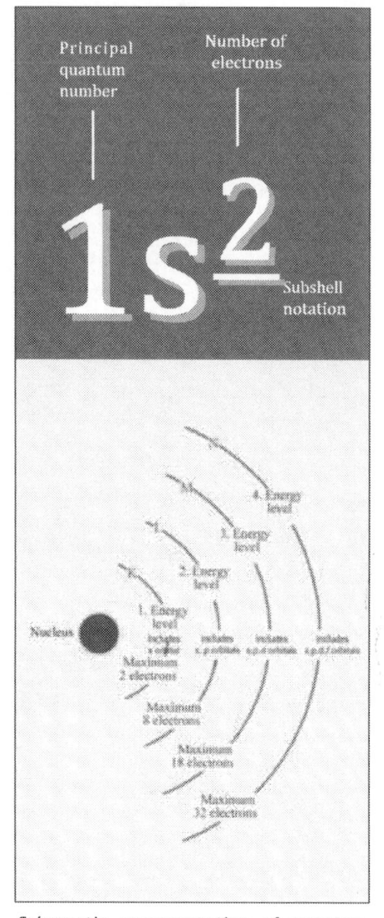

Schematic representation of quantum numbers and electron distribution.

The principal quantum number, n

- determines the size and energy of an atom (larger n = bigger atoms and higher energy)
- can take an integer value n = 1, 2, 3, 4 ... (K, L, M, N...)
- all electrons in an atom with the same value are said to belong to the same shell

The secondary quantum number, l

- determines the overall shape of the orbital within a shell
- affects orbital energies (bigger l = higher energy)
- all electrons in an atom with the same value of 'l' are said to belong to the same subshell
- has integer values between 0 and n-1
- may be called the 'orbital angular momentum quantum number'

The magnetic quantum number, m_l

- determines the orientation of orbitals within a subshell
- does not affect orbital energy
- has integer values between -l and +l
- the number of m_l values within a subshell is the number of orbitals within a subshell
- s, p, d and f subshells includes 1, 3, 5 and 7 orbitals respectively

The spin quantum number, m_s

- each orbital may contain two electrons at most
- several experimental observations can be explained by treating the electron as though it were spinning
- spin affects the electron behave like a tiny magnet
- spin can be clockwise (+1/2) or counterclockwise (-1/2)

The table below shows the quantum numbers of electrons in the first four shells. You can find more information about quantum numbers in Appendix B.

Principal quantum number (n)	K					M									N															
Secondary quantum number (l)																														
Magnetic quantum number (m_l)																-1														+3
Magnetic spin quantum number (m_s) ↑: +1/2	-1/2	-1/2	-1/2	-1/2	-1/2	+1/2	+1/2	+1/2	+1/2	+1/2	+1/2	+1/2	+1/2	-1/2	-1/2	-1/2	-1/2	-1/2	-1/2	-1/2	-1/2	-1/2	-1/2	-1/2	-1/2	-1/2	-1/2	-1/2	-1/2	-1/2
↓: -1/2	-1/2	-1/2	-1/2	-1/2	-1/2	-1/2	-1/2	-1/2	-1/2	-1/2	-1/2	-1/2	-1/2	-1/2	-1/2	-1/2	-1/2	-1/2	-1/2	-1/2	-1/2	-1/2	-1/2	-1/2	-1/2	-1/2	-1/2	-1/2	-1/2	-1/2
																														↑↓
Electron capacity: Orbital	2	2	2	2	2	2	2	2	2	2	2	2	2	2	2	2	2	2	2	2	2	2	2	2	2	2	2	2	2	2
Electron capacity: Subshell	2	2	6			2	6			10					2	6			10					14						
Electron capacity: Orbit	2	8				18									32															

The set of four quantum numbers (n, l, m_l and m_s) for an electron in the first four energy levels (K, L, M and N) is shown. The K – level, which is the closest shell to the nucleus, can hold a maximum of two electrons. These electrons occupy the 1 – s orbital (for which l = 0). The L – level, which is the second closest energy level to the nucleus, can hold a maximum of 8 electrons. These electrons occupy the 2s – orbital (l= 0), and 2p – orbitals (l = 1) consecutively. The M – level, which is the third nearest energy level, can hold a maximum of 18 electrons. These electrons occupy the 3s - orbital (l = 0), 3p - orbitals (l = 1) and 3d - orbitals (l = 2) respectively. The N – level, the furthest shell, can hold a maximum of 32 electrons. These electrons occupy the 4s - orbital (l = 0), 4p - orbitals (l = 1), 4d - orbitals (l = 2) and 4f - orbitals (l = 3) consecutively.

Summary : History of Atom with Scientists

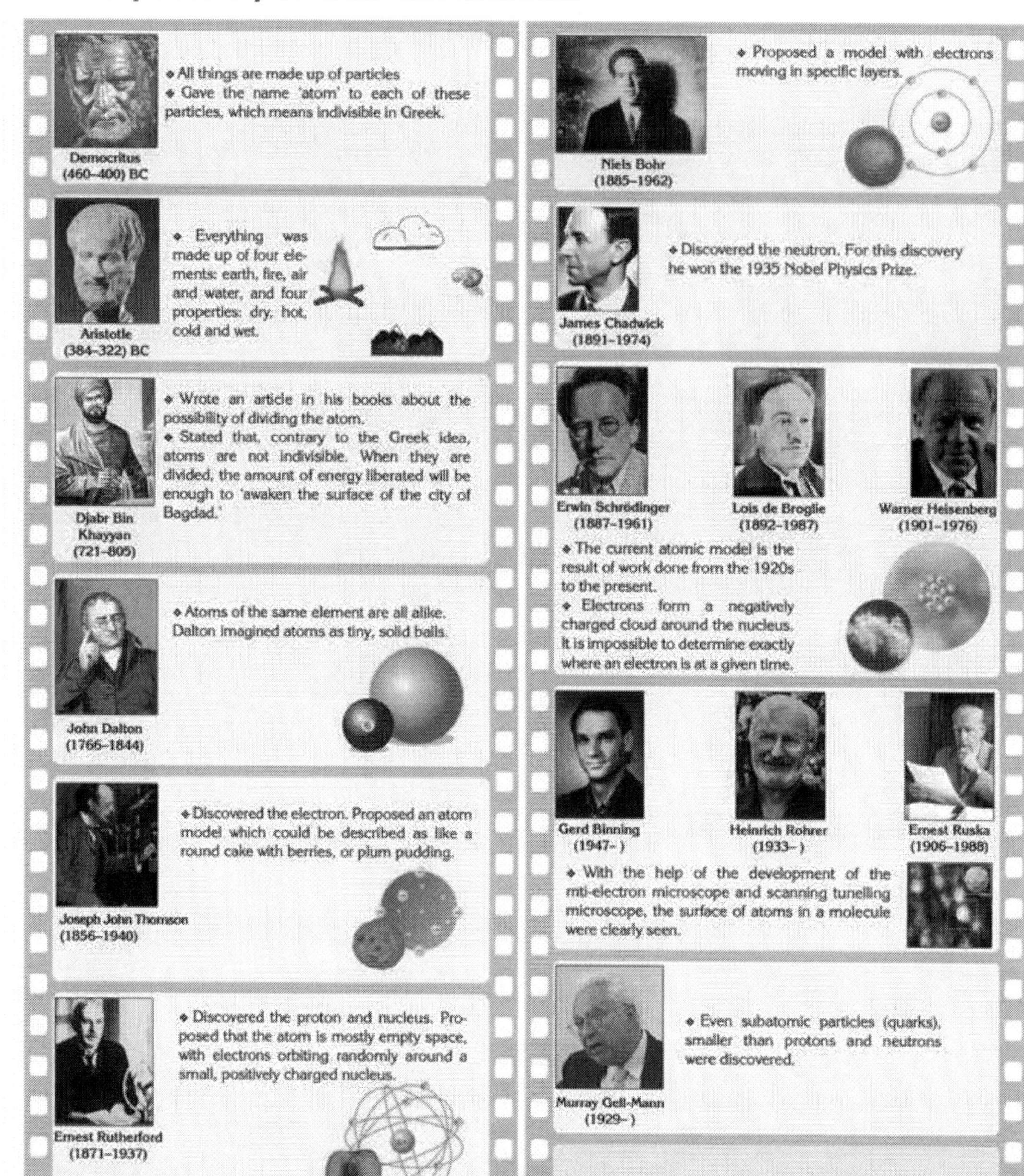

2.2. ELECTRON CONFIGURATION

Rules of Electron Distribution

Pauli's Principle

The electron of a hydrogen atom in its ground state is located in the nearest orbital to the nucleus. But, what about the electron distribution of the atoms with more than one electron? Answering this question in 1925, Wolfgang Pauli stated his exclusion principle thus:

'In the same atom, two electrons may not have identical sets of all quantum numbers.'

According to this principle, the quantum numbers, n, l, m_l, and m_s, can never be identical for two electrons in an atom. This means that at least one of the quantum numbers must be different. For example, even if two electrons have identical values for n, l and m_l (as a result of being in the same orbital), their magnetic spin quantum numbers must be different. That is, these electrons are said to have opposing spins. In fact, we have already mentioned that each electron may be described by a set of the four quantum numbers;

- n shows the shell and the relative average distance of the electron from the nucleus
- l shows the subshell and the shape of the orbital for the electron
- m_l represents the orientation of the orbital in spaces
- m_s refers to the spin of the electron.

The Aufbau Principle

The Aufbau principle basically states that the lowest energy orbitals are filled first. 1s orbital has the lowest energy, so it is first to be filled, followed, in order, by 2s, 2p, 3s... This ordering was first stated by Wolfgang Pauli and is called the Aufbau principle (aufbau means 'building up' in German).

Hund's Rule

Different orbitals with identical energy (those in the same subshell) are known as equal energetic orbitals. For example the orbitals of p subshell p_x, p_y and p_z are of identical energy. Since all electrons carry the same electrical charge, they tend to be as far as possible from each other. Thus, Hund's rule states that the electrons are distributed among the orbitals of a subshell of the same energy in a way that gives the maximum number of unpaired electrons with parallel spin. The term 'parallel spin' means that all the unpaired electrons spin in the same direction, and all of the m_s values of these electrons have the same sign.

In some cases, at higher quantum levels, since the energies of some subshells are very close to each other, there may not be any coherence between the order

THE PIONEERS

Wolfgang Pauli (1900 - 1958)

Pauli, an Austrian physicist, is best known for his exclusion principle regarding the distribution of electrons among the atomic orbitals, for which he was awarded the 1945 Nobel Physics Prize. In his exclusion principle, Pauli says that no two electrons in the same atom may have identical sets of all four quantum numbers.

Ground state: *The state in which all the electrons in an atom are in the lowest energy levels available. For example,* $_{12}Mg: 1s^2\ 2s^2\ 2p^6\ 3s^2$

Excited state electron configuration: *When an atom has absorbed energy, its electrons may move to higher state energy levels. For example,* $_{12}Mg: 1s^2\ 2s^2\ 2p^6\ 3s^1\ 3p^1$

Aufbau principle (Aufbau process) is also called Aufbau order.

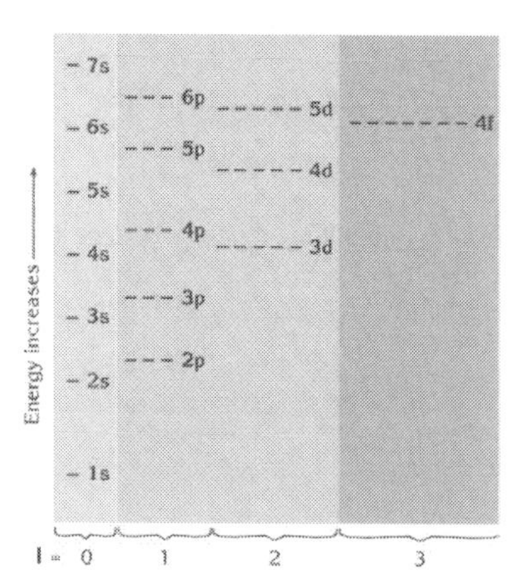

General order of filling of the orbitals in an atom.

of filling the orbitals and the order of increasing the energies of the orbitals. For example, the 4s orbital has a lower energy than that of a 3d orbital. That is, the 4s orbital is filled before the 3d orbital. The order of filling of orbitals is derived as a result of experiments in spectroscopy and magnetism. The order of filling electrons in atomic orbitals, with a few exceptions, is roughly as follows.

$1s^2$, $2s^2$, $2p^6$, $3s^2$, $3p^6$, $4s^2$, $3d^{10}$, $4p^6$, $5s^2$, $4d^{10}$, $5p^6$, $6s^2$, $4f^{14}$, $5d^{10}$, $6p^6$, $7s^2$, $5f^{14}$, $6d^{10}$, $7p^6$

In order to remember and derive this order easily, the method illustrated by the table below left is very useful. When following the arrows from top to bottom, the order given above is obtained.

Representation of Electron Configuration

The electron configuration of an atom can be represented by either electronic notation (s, p, d, f) or by orbital diagram.

For instance, the electron configuration of the silicon, Si, atoms for which the atomic number is 14 (that is, the number of electrons is 14) is given below.

Si: $1s^2\ 2s^2\ 2p^6\ 3s^2\ 3p^2$

Orbital diagram notation :

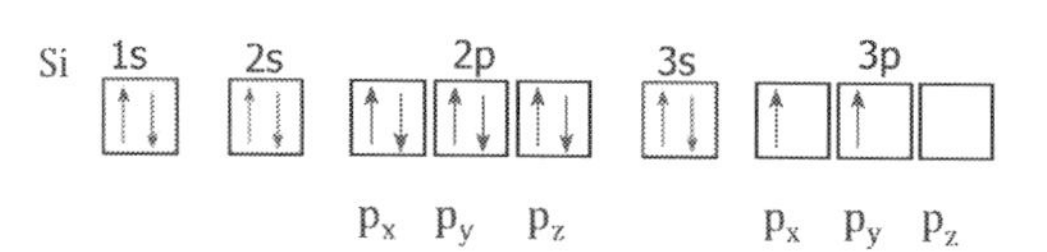

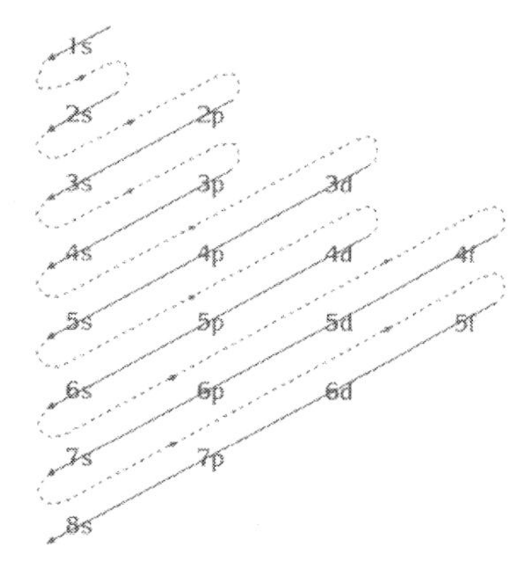

The Aufbau order of filling the atomic orbitals.

In the orbital diagram notation, each subshell is divided into individual orbitals drawn as boxes. An arrow pointing upward corresponds to one type of spin (+1/2) and an arrow pointing down corresponds to the opposite spin (-1/2). Electrons in the same orbital with opposed spins are said to be paired, such as the electrons in the 1s and 2s orbitals. These orbitals are completely filled orbitals.

On the other hand, since electrons are placed one by one in a subshell with parallel spins, the corresponding arrows are drawn in the same direction, such as in the 3p electrons of the silicon atom. Such orbitals are half-filled orbitals.

Electron Configuration of the Elements

Now, let us give some systematic examples to illustrate how electrons are placed in the orbitals.

For the hydrogen atom (Z = 1): since the least possible energy level of an electron is 1s, the single electron of hydrogen occupies the first energy level. So, the corresponding electron distribution is $1s^1$.

$_{1}H\ 1s^1$

For the helium atom (Z = 2): the first electron of helium atom will be distributed just like the single hydrogen electron, but the second electron of the helium atom will be placed into the 1s orbital with an opposite spin. So, the corresponding electron distribution is $1s^2$.

$_{2}He\ 1s^2$

For the lithium atom (Z = 3): since the $1s^2$ orbital has already been occupied by the first two electrons, the third electron is added to the 2s orbital. So, the corresponding electron distribution is $1s^2\ 2s^1$.

$_{3}Li\ 1s^2\ 2s^1$

For the sodium atom (Z=11): through the argon atom (Z=18) (Na, Mg, Al, Si, P, S, Cl and Ar); the electrons occupy their 3s orbital, then the 3p orbitals.

$_{11}Na\ 1s^2\ 2s^2\ 2p^6\ 3s^1$

You can find the electron configurations of all the elements in Appendix A.

The corresponding electron distributions of these elements are given below.

$_{11}Na$: $1s^2\ 2s^2\ 2p^6\ 3s^1$
$_{12}Mg$: $1s^2\ 2s^2\ 2p^6\ 3s^2$
$_{13}Al$: $1s^2\ 2s^2\ 2p^6\ 3s^2\ 3p^1$
$_{14}Si$: $1s^2\ 2s^2\ 2p^6\ 3s^2\ 3p^2$
$_{15}P$: $1s^2\ 2s^2\ 2p^6\ 3s^2\ 3p^3$
$_{16}S$: $1s^2\ 2s^2\ 2p^6\ 3s^2\ 3p^4$
$_{17}Cl$: $1s^2\ 2s^2\ 2p^6\ 3s^2\ 3p^5$
$_{18}Ar$: $1s^2\ 2s^2\ 2p^6\ 3s^2\ 3p^6$

If you pay attention, it is easy to see that the 1s, 2s and 2p orbitals of these elements are full. Here the electron distribution of $1s^2 2s^2 2p^6$ resembles the electron distribution of neon. This is generally denoted by [Ne].

In the case of writing the electron distribution of the elements after neon, we can start with the [Ne] symbol and then write the electrons in the last (outermost) shell.

These electrons, which have the biggest principal quantum number, are called 'valence' electrons. Thus we may rewrite the electron configuration of these elements as follows:

Na : [Ne] $3s^1$
Mg : [Ne] $3s^2$
Al : [Ne] $3s^2\ 3p^1$
Si : [Ne] $3s^2\ 3p^2$
P : [Ne] $3s^2\ 3p^3$
S : [Ne] $3s^2\ 3p^4$
Cl : [Ne] $3s^2\ 3p^5$
Ar : [Ne] $3s^2\ 3p^6$

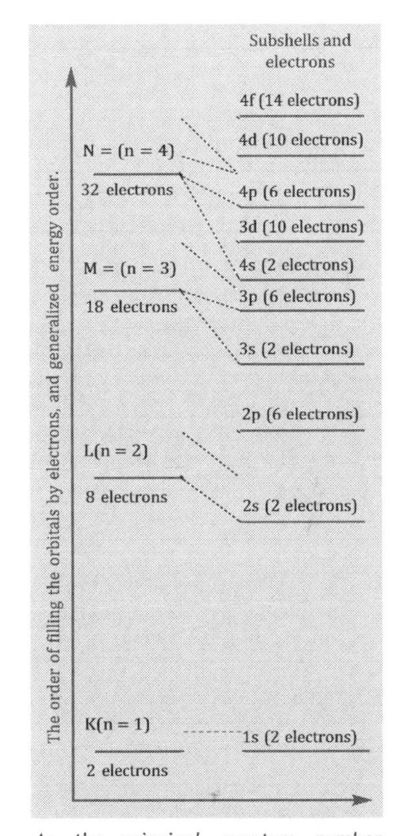

As the principal quantum number increases, the energies of the electrons in the orbitals generally increase. However, as seen from the table above, the valence electrons of potassium and calcium tend to occupy 4s - orbitals rather than 3d - orbitals. Hence, the energies of 4s - orbitals (n = 4) are smaller than 3d - orbitals.

SUPPLEMENTARY QUESTIONS

1. What was the basic philosophy of Aristotle on the components of matter?

2. How did Democritus define the atom? How are Democritus' ideas still valid today?

3. Who first put the atomic concept on a scientific basis? Explain.

4. How are Aristotle's opinions about matter still valid today?

5. What do the atomic models of Dalton, Thomson, Rutherford and Bohr look like?

6. Who discovered the proton, the neutron and the electron?

7. Who are the pioneers of modern atomic theory?

8. Which scientists were awarded of the Nobel Prize in Physics for their visualization of the carbon atom using the Scanning Tunelling Microscope (STM)?

9. When Dalton was establishing his own atomic model, how did he utilize the old knowledge? What was Dalton's atomic model like?

10. Which aspects of Dalton's atomic model are deficient or false with respect to our today's understandings?

11. Who first discovered the electron? Explain.

12. What are the factors affecting the movement of electrons in an electric field? In which direction do the electrons deviate in an electric field and why?

13. How could you prove that electrons are fundamental particles present in every substance?

14. What are the basic differences between the proton and the electron?

15. Explain Thomson's atomic model. Which aspects of Thomson's model are deficient or false with respect to our modern understanding?

16. Summarize the Rutherford's α – particle experiment. What is the most important result that he derived from his experiment?

17. Summarize Rutherford's atomic model. Which aspects of this model are deficient or false as far as today's understanding is concerned?

18. Answer the following questions about Bohr's atomic model.
    - **a.** What was the main deficiency of Bohr's time atomic models that forced him to do investigations?
    - **b.** What are the modern terms contributed by Bohr's atomic model to the history of science which are still used today?
    - **c.** According to Bohr's opinion, when does an atom emit light?
    - **d.** Compare Bohr's atomic model with modern atomic model.

19. Define the following terms and concepts

a. Photoelectric event,

b. Heisenberg's uncertainty principle

c. Hund's rule

d. The Aufbau principle

20. What are the concepts about Bohr's model which was not enough to explain?

21. As far as Heisenberg's uncertainty principle is concerned, why is it impossible to determine the exact position and momentum of an electron simultaneously?

22. Describe the modern atomic model.

23. License plates are used to distinguish cars in a city, even in the world. Can you establish a similar relationship between the four quantum numbers and electrons? How?

24. Why cannot two electrons of an atom have the same sets of four quantum numbers?

25. Define the following terms.

a. Principal quantum numbers

b. Secondary quantum numbers

c. Magnetic quantum numbers

d. Spin quantum numbers

26. Compare the energies of the sublevels of s, p, d and f found in the same energy level.

27. Fill in the blanks in the table below.

	Number of l	Orbital representation
n = 2	1	
n = 3	0	
n = 4	3	
n = 5	2	

28. Write the orbital representation for each of the following l values.

a. l = 0 **b.** l = 1 **c.** l = 2 **d.** l = 3

29. Write the electron configuration and find the maximum numbers of electrons that can be present in each of the following energy levels.

a. In n = 4 l = 1

b. In n = 3 l = 1 and l = 2

c. In n = 4 l = 0 and l = 3

30. How many electrons can be present in n = 4, l = 3, m_l = +3 of element uranium (U), and in which sublevel is this orbital?

31. What is the maximum number of electrons present in n = 3 having the value of m_s = + 1/2 ?

32. Find the values of the quantum numbers missing in each of the following situations.

a. n = 4 l = 0 m_l = ?

b. n = 3 l = 1 m_l = ?

c. n = 3 l =? m_l = −1

d. n = ? l = 1 m_l = +1

33. Find the appropriate values of n and l for each of the following subshells.

a. 3s **b.** 4p **c.** 5f **d.** 3d

34. What is the maximum number of orbitals present in each of the following subshells?

a. 2s **b.** 3d **c.** 4p **d.** 4f **e.** 5d

35. Write down the electron configurations of the following elements.

a. $_{11}Na$ **b.** $_{13}Al$ **c.** $_{15}P$ **d.** $_{17}Cl$ **e.** $_{18}Ar$

f. $_{19}K$ **g.** $_{20}Ca$ **h.** $_{22}Ti$ **i.** $_{23}V$ **j.** $_{24}Cr$

k. $_{25}Mn$ **l.** $_{30}Zn$ **m.** $_{31}Ga$ **n.** $_{36}Kr$ **o.** $_{56}Ba$

36. Write down the electron configuration of the following ions.

a. $_{20}Ca^{2+}$ **b.** $_{30}Zn^{2+}$ **c.** $_{25}Mn^{2+}$

d. $_{26}Fe^{3+}$ **e.** $_{17}Cl^{-}$ **f.** $_{34}Se^{2-}$

MULTIPLE CHOICE QUESTIONS

1. Which famous scientist said, 'We can divide matter into infinitely many subdivisions and we still would not obtain an indivisible piece of matter'?

A) Democritus B) Aristotle C) Khayyan D) Mendeleev E) Lavoisier

2. Who put the atomic concept on a scientific basis?

A) Aristotle B) Dalton C) Bohr D) Mendeleev E) Lavoisier

3. In which option are the atomic models of Dalton, Thomson, Rutherford and Bohr are given in the correct order?

A) berries in cake, hard ball, onion, peach
B) peach, berries in cake, hard ball, onion
C) hard ball, berries in cake, peach, onion
D) onion, peach, berries in cake, hard ball
E) peach, onion, hard ball, berries in cake

4. Who first talked about the huge amount of energy of an atom?

A) J. J. Thomson
B) Aristotle
C) Plato
D) Djabr Bin Khayyan
E) John Dalton

5. In which option is the development of atomic models is given in the correct order?

A) Thomson, Rutherford, Bohr, Dalton
B) Dalton, Thomson, Bohr, Rutherford
C) Dalton, Thomson, Rutherford, Bohr
D) Thomson, Bohr, Rutherford, Dalton
E) Bohr, Dalton, Rutherford, Thomson

6. In which option are the discoverers of the proton, neutron and electron given in the correct order?

A) Thomson, Dalton, Rutherford
B) Rutherford, Chadwick, Thomson
C) Moseley, Dalton, Thomson
D) Mendeleev, Moseley, Einstein
E) Einstein, Chadwick, Thomson

7. Which one of the following atomic theories proposed by Dalton is wrong according to modern atomic theory?

A) Atom shows all of the characteristic properties of its substances.
B) Atoms of an element are completely identical.
C) Atoms are filled spheres.
D) Different type of atoms have the different masses.
E) The smallest building bricks of substances are atoms.

8. In which option are the pioneers of modern atomic theory is given in the correct order?

A) Broglie, Heisenberg, Schrödinger
B) Einstein, Heisenberg, Schrödinger
C) Broglie, Thomson, Dalton
D) Schrödinger, Dalton, Thomson
E) Rutherford, Broglie, Heisenberg

9. The modern atomic theory depends on which information below?

A) The charge of subatomic particles
B) Wave mechanics to the movement of electrons.
C) The possible stationary energy levels of electrons.
D) The distribution of protons and electrons throughout the atom.
E) The mass of electrons.

10. The elements C and H form more than one compound. The graph opposite gives the relationship between the masses of the constituents, C and H, and the masses of the both compounds formed. If the molecular formula of the first compound is CH_4, what is the empirical formula of the second compound?

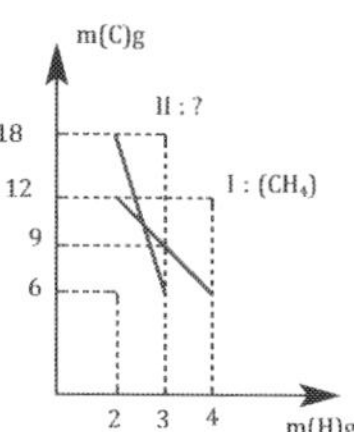

A) CH_3 B) C_2H_2 C) C_3H_4 D) CH_2 E) CH_4

11. Which of the following statements explains Dalton's atomic model correctly?

A) An atom is a sphere, and at the centre of this sphere there is a nucleus.
B) Atoms are filled spheres.
C) An atom is composed of a nucleus, around which the electrons revolve.
D) Protons are distributed randomly among the surface of the atoms.
E) Electrons move randomly around the nucleus.

12. Which one of the following statements is not the mistake of Dalton's atomic model?

A) Atoms are empty structures.
B) The atoms of the same type elements are totally identical.
C) It is not true that atoms are indivisible.
D) Atoms form compounds by combining with a definite number ratio.
E) Atoms show the characteristics of that element.

13. Which one of te following electron structures belongs to the ion $_{12}Mg^{2+}$?

A) $[Ne]3s^1$ B) $[Ne]3s^2$ C) [Ne]
D) $1s^2\ 2s^2\ 2p^4$ E) $1s^2\ 2s^2\ 2p^5$

14. Which one of the following electron distributions does not follow Hund's rule?

A) $1s^2\ 2s^2\ 2p^2_x\ 2p^1_y\ 2p^1_z$ B) $1s^2\ 2s^2\ 2p^1_x\ 2p^1_y$
C) $1s^2\ 2s^2\ 2p^1_x\ 2p^1_y\ 2p^1_z$ D) $1s^2\ 2s^2\ 2p^2_x\ 2p^1_y\ 2p^0_z$
E) $1s^2\ 2s^2\ 2p^2_x\ 2p^2_y\ 2p^1_z$

15. What is the atomic number of an element having a total of 5 electrons in n = 4?

A) 24 B) 28 C) 30 D) 33 E) 35

16. In a compound X_2O_3, the ratio of the mass of X to the mass of compound is 7/19. If the atom of element X has a total of 7 neutrons in its nucleus, what is the electron configuration of X?

A) $_9X:1s^2\ 2s^2\ 2p^5$ B) $_8X:1s^2\ 2s^2\ 2p^4$
C) $_7X:1s^2\ 2s^2\ 2p^3$ D) $_6X:1s^2\ 2s^2\ 2p^2$
E) $_5X:1s^2\ 2s^2\ 2p^1$

17. Which one of the following electron distributions does not follow the Aufbau principle?

A) $_9F:1s^2\ 2s^2\ 2p^5$ B) $_8O:1s^2\ 2s^2\ 2p^4$
C) $_7N:1s^2\ 2s^2\ 2p^3$ D) $_6C:1s^2\ 2s^1\ 3s^2\ 3p^1$
E) $_5B:1s^2\ 2s^2\ 2p^1$

18. What is the maximum number of electrons in n = 3 with the m_s value of + 1/2?

A) 9 B) 8 C) 7 D) 6 E) 5

19. How many electrons can be present in n=2 with an m_l value of 0?

A) 4 B) 5 C) 6 D) 8 E) 10

20. How many electrons can be present in n=4 with an l value of 2?

A) 10 B) 8 C) 6 D) 4 E) 2

21. For the electron configuration, X: $1s^2\ 2s^2\ 2p^6\ 3s^2\ 3p^1\ 4s^1$, find the set of four quantum numbers n, l, m_l and m_s respectively for the valence electrons of the X atom.

A) 3, 2, +1, +1/2 B) 3, 2, +1, +1/2
C) 2, 1, +1, -1/2 D) 3, 1, -1, +1/2
E) 4, 0, 0, +1/2

22. For the electron configuration, X: $[Ne]3s^2\ 3p^1$, find the set of four quantum numbers n, l, m_l and m_s respectively for the outer most electron of the X atom.

A) 3, 1, +1, +1/2 B) 3, 2, +1, +1/2
C) 3, 0, +1, -1/2 D) 3, 1, -1, +1/2
E) 3, 1, 0, +1/2

23. Which one of the following configurations is the electron configuration of an element whose valence electron has the set of four quantum numbers of;

n=4, l =1, m_l =-1 and m_s =+1/2?

A) X: $1s^2\ 2s^2\ 2p^6\ 3s^2\ 3p^6\ 4s^1$
B) X: $1s^2\ 2s^2\ 2p^6\ 3s^2\ 3p^6\ 4s^2\ 4p^3$
C) X: $1s^2\ 2s^2\ 2p^6\ 3s^2\ 3p^6\ 4s^2\ 4p^1$
D) X: $1s^2\ 2s^2\ 2p^6\ 3s^2\ 3p^6\ 4s^2$
E) X: $1s^2\ 2s^2\ 2p^6\ 3s^2\ 3p^6\ 4s^2\ 4p^4$

CROSSWORD PUZZLE

ACROSS

5 Greek philosopher who stated matter was continuous, there was no limit on how finely you could cut it up.

8 Scientist who discovered the electron.

9 Subatomic particle that has no charge.

11 Scientist who discovered the neutrons.

12 Part of nucleus.

13 Scientist who discovered the protons and nucleus.

DOWN

1 Imagined atoms as tiny, solid balls.

2 Scientist who proposed the model that electrons moving in specific layers.

3 Negatively charged subatomic particle.

4 Positively charged subatomic particle.

6 Greek philosopher who named the smallest partical an atom.

7 Scientist who proposed that two electrons in the same atom do not have identical sets of all four quantum numbers.

10 Scientist who first talked about the huge amount of energy of the atom.

WORD SEARCH PUZZLE

Find the words in the grid. Words can go horizontally, vertically and diagonally in all eight directions.

X	B	M	▪	T	N	A	▪	Q	T	X	Q	N	M	N
R	B	S	C	H	R	O	D	I	N	G	I	R	K	C
L	R	P	N	O	R	T	▪	I	N	T	H	N	R	Y
R	O	R	B	I	T	A	L	N	R	I	T	▪	A	€
C	G	O	Q	Q	N	A	O	R	I	O	T	X	▪	W
H	L	T	€	C	N	S	R	S	W	H	H	S	Q	I
A	I	O	V	D	M	N	I	I	I	S	▪	B	L	L
D	I	N	O	O	Q	N	U	R	S	T	H	€	R	▪
W	M	Q	H	R	B	B	F	N	I	T	D	I	T	A
I	O	T	€	I	B	O	C	R	O	U	O	€	L	P
C	T	G	R	€	R	I	C	N	A	T	W	T	K	L
K	A	G	N	D	H	O	T	B	Y	N	L	W	L	B
G	€	I	T	K	M	P	R	R	R	W	D	A	H	I
K	P	T	F	I	L	I	C	T	R	O	N	G	D	T
S	T	N	D	B	H	B	M	H	B	L	C	U	F	X

NEUTRON	QUARK	ATOM	DJABR
ORBIT	RUTHERFORD	BOHR	ELECTRON
ORBITAL	SCHRODINGER	BROGLIE	HEISENBERG
PAULI	SHELL	CHADWICK	THOMSON
PROTON	SPIN	DALTON	
QUANTUM	ARISTOTLE	DEMOCRITUS	

NOTES

THE PERIODIC TABLE

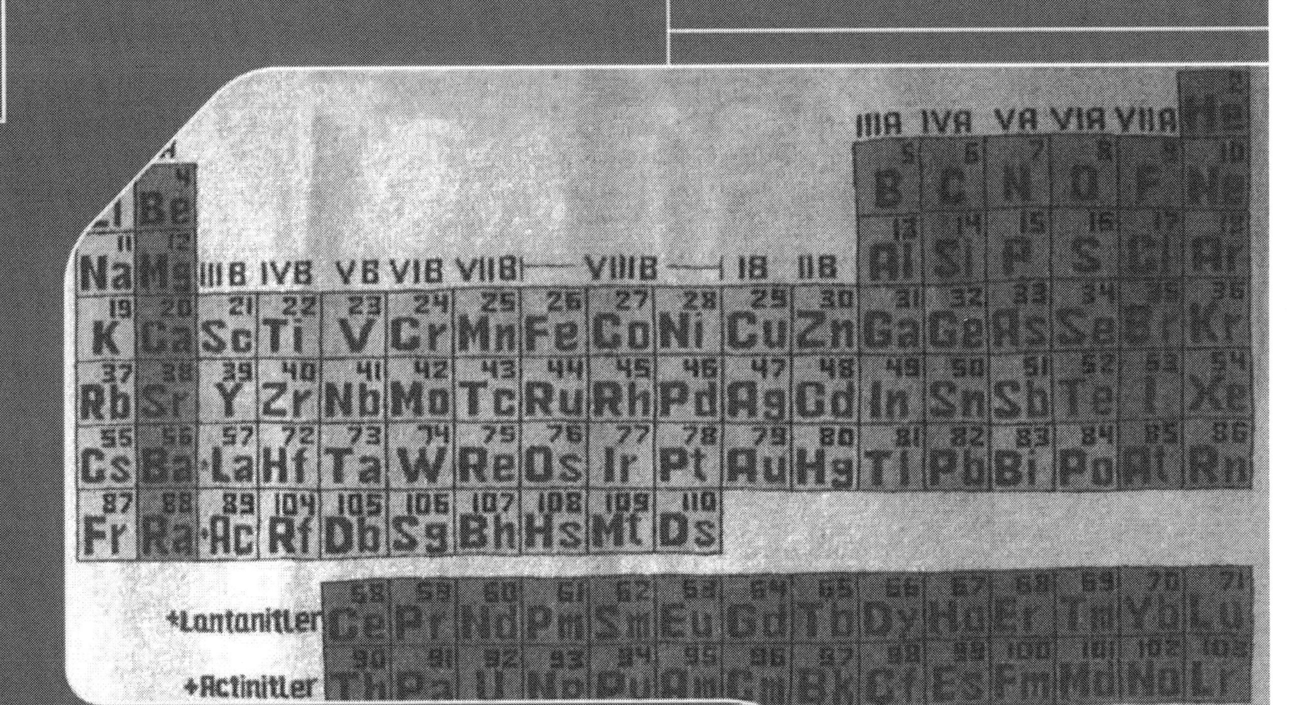

Mendeleev is often considered the father of the periodic table.

Naming Elements

There no fixed rule for naming elements. Some names of elements take the names of scientists (Mendelevium, Md; Rontgenium, Rg...), places (Americium, Am; Europium, Eu...), and the names of planets (Uranium, U; Plutonium, Pu...). Some names come from mythology (Titanium, Ti; Thorium, Th...) or properties of the element (eg hydrogen means water former, oxygen means acid former).

Some elements have been known since ancient times (eg Iron, Fe; Tin, Sn). Today, some artifical elements are waiting to be named in the periodic table (eg Ununbium, Uub, Ununquadium, Uuq).

INTRODUCTION

Dmitri Mendeleev is often considered the "father" of the periodic table, and the studies of many scientists have also contributed to form the modern periodic table.

The discovery of individual elements was necessary to begin to the construction of the periodic table.

Although the elements, gold (Au), silver (Ag), tin (Sn), copper (Cu), lead (Pb), mercury (Hg), antimony (Sb), arsenic (As), carbon (C), iron (Fe), sulfur (S) and zinc (Zn) had been known since ancient times, the first scientific discovery of an element occurred in 1669, when Hennig Brand discovered phosphorus (P). During the next two hundred years various chemical and physical properties of elements and their compounds were studied by chemists. By 1869, a total of 63 elements had been discovered. As the number of known elements increased, scientists began to recognize patterns in properties and began to develop classification schemes.

In 1869 Mendeleev and Meyer independently proposed the periodic law. The periodic law states that when the elements are arranged in order of increasing atomic mass, certain sets of properties occur periodically.

The periodic table prepared by Mendeleev became a target for scientists and elements as the blank spaces in his table were quickly discovered. Of great importance also was the indication of the presence of the trans-uranium elements in Mendeleev's studies.

Since Mendeleev arranged the elements according to their atomic masses, he could not be aware of the lack of noble gases in his table. After the discovery of the noble gases by William Ramsay in the years between 1894 and 1900, an English physicist named Henry Moseley solved the puzzle of the periodic table in 1913.

Moseley determined the number of protons (atomic numbers) of the elements by the wavelength of X-rays emitted by atoms which were bombarded by electrons. He claimed that the elements in the periodic table must be arranged according to their atomic numbers instead of their atomic masses. Thus, the modern periodic table appeared.

"The element coming after any consecutive eight elements is somehow repetition of the first element like sound of an octave in music."

John A. R. NEWLANDS 1864

READING

Mendeleev's Periodic Table

In 1869, two scientists working independently and unaware of each other, Dmitri Mendeleev (a Russian chemist) and Lothar Meyer (a German scientist) made similar classifications of the elements. Both scientists classified the elements in the order of increasing atomic mass, and, as a result, they noticed some similar periodic properties among some elements. Mendeleev's work and ideas on periodic properties of elements attracted much attention.

Mendeleev arranged 63 elements horizontally in the order of increasing atomic mass in his periodic table. This table was composed of 12 horizontal rows and 8 vertical columns. Mendeleev gave formulas of the elements in each group with hydrogen, oxygen and chlorine atoms at the top of his table. He found out that the physical and chemical properties of the elements in a vertical column are similar and these similarities change regularly within a group. With this discovery, Mendeleev introduced the concept of 'families' of elements. Elements having similar chemical properties were placed in the same vertical column.

He purposefully left some blank places in his table and said that the elements corresponding to these spaces would be discovered. Mendeleev guessed some elements later to be placed into these spaces, for example, the element gallium, which was not known in 1869.

ROW	Group 1	Group 2	Group 3	Group 4	Group 5	Group 6	Group 7	Group 8
	R_2O	RO	R_2O_3	RO_2	R_2O_5	RO_3	R_2O_7	RO_4
	RCl	RCl_2	RCl_3	RCl_4 RH_4	RH_3	RH_2	RH	
1	H							
2	Li	Be	B	C	N	O	F	
3	Na	Mg	Al	Si	P	S	Cl	
4	K	Ca		Ti	V	Cr	Mn	Fe, Co, Ni, Cu
5	Cu	Zn			As	Se	Br	
6	Rb	Sr	Yt	Zr	Nb	Mo		Ru, Rh, Pd, Ag
7	Ag	Cd	In	Sn	Sb	Te	I	
8	Cs	Ba	Di	Ce				
9								
10			Er	La	Ta	W		Os, Ir, Pt, Au
11	Au	Hg	Tl	Pb	Bi			
12				Th		U		

* R represents the group elements in the formulas.

The final form of Mendeleev's periodic table prepared by Mendeleev in 1871.

Elements and Discovery Date	Properties	Predictions of Mendeleev in 1871	Determined Values
Eka-Aluminum, Ea (Gallium, Ga) 1875	Atomic mass	68 g/mol	69.7 g/mol
	Density	6 g/mL	5.96 g/mL
	Melting point	Low	30°C
	Formula of its oxide	Ea_2O_3	Ga_2O_3
	Solubility of its oxide	Soluble in ammonia solution	Soluble in ammonia solution

Estimated predictions of Mendeleev about gallium and its determined values.

Mendeleev had determined some physical and chemical properties of elements before they were discovered and gave these unknown elements names such as eka-aluminum, eka-silicon, eka-boron, eka-cesium andeka-iodine.

The term "eka" is derived from sanskrit and means "first".

You can find the updated Mendeleev's periodic table of in Appendix D.

Although cobalt is heavier than nickel, it is placed before nickel in contrast to Mendeleev's periodic rule. He thought that reason of this was the incorrect measurement of the atomic masses.

DEVELOPMENT OF THE PERIODIC TABLE		
Scientist	**Date**	**Contribution**
Antoine Lavoisier	1770 – 1789	Wrote the first extensive list of elements containing 33 elements. Distinguished between metals and nonmetals. Some of Lavoisier's elements were later shown to be compounds and mixtures.
Jöns Jakob Berzelius	1828	Developed a table of atomic weights. Introduced letters to symbolize elements.
Johann Döbereiner	1829	Developed "triads", groups of three elements with similar properties. Lithium, sodium and potassium formed a triad. Calcium, strontium and barium formed a triad. Chlorine, bromine and iodine formed a triad. For example the properties of bromine are between the chlorine and iodine.
John Newlands	1864	Arranged the known elements (>60) in order of atomic weights and observed similarities between the first and ninth elements, the second and tenth elements etc. He proposed the 'Law of Octaves'.
Lothar Meyer	1869	Complied a periodic table of 56 elements based on the periodicity of properties such as molar volume when arranged in order of atomic weight. Meyer and Mendeleev produced their periodic tables simultaneously.
Dmitri Mendeleev	1869	Produced a table based on atomic weights but arranged 'periodically' with elements with similar properties under each other. Left gaps for elements that were unknown at that time and predicted their properties (the elements were gallium, scandium and germanium). Rearranged the order of elements if their properties dictated it, eg, tellurium is heavier than iodine but comes before it in the periodic table.
William Ramsay	1894	Discovered the Noble Gases. In 1894 Ramsay removed oxygen, nitrogen, water and carbon dioxide from a sample of air and was left with a gas 19 times heavier than hydrogen, very unreactive and with an unknown emission spectrum. He called this gas as argon. In 1895 he discovered helium as a decay product of uranium and matched it to the emission spectrum of an unknown element in the sun that was discovered in 1868. He went on to discover neon, krypton and xenon, and realized these represented a new group in the periodic table.
Henry Moseley	1919	Determined the atomic number of each of the elements. Modified the "Periodic Law" to read that the properties of the elements vary periodically with their atomic numbers.
Glenn Seaborg	1940	Synthesized transuranic elements (the elements after uranium in the periodic table). In 1940, bombarded uranium with neutrons in a cyclotron to produce neptunium (Z = 93). Produced plutonium (Z = 94) from uranium and deuterium. These new elements were part of a new block of the periodic table called Actinides.

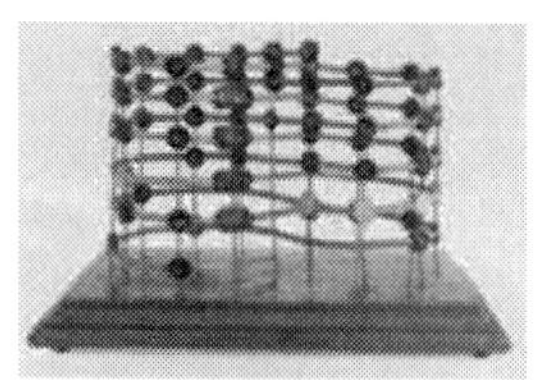

Sir William Crooks

Dmitri Mendeleev

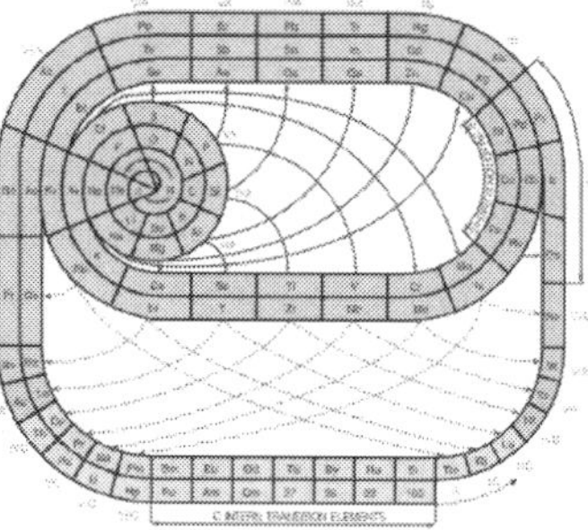

Charles Janet

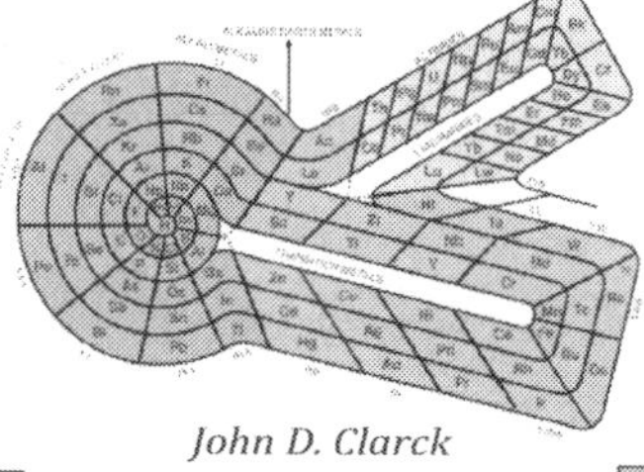

John D. Clarck

A modern periodic table.

Some periodic tables.

1. THE MODERN PERIODIC TABLE

The modern periodic table appeared as a function of the physical and chemical properties of elements. When the elements are arranged in the order of increasing atomic numbers, there is a periodic repetition in the properties of these elements.

A simple periodic table contains the symbols, atomic numbers and the relative atomic masses of the elements. Additionally, detailed periodic tables containing some physical and chemical properties (such as melting point, boiling point, oxidation state) are also made.

Each horizontal row in the periodic table is called a **period**. There are seven periods in the modern periodic table and each period begins with a metal and ends with a noble gas. However, the first element of the first period (hydrogen) is not a metal. Additionally, the noble gas of the seventh period has not been discovered yet.

Each vertical column in the periodic table is called a **group**. Since the chemical and physical properties of the elements in a group are similar, they are sometimes also called a **family**. There are a total of eighteen groups in the periodic table of which eight are A groups and eight are B groups. The group 8B contains three columns. A groups are called **main groups (representative group)** and B groups are called **transition metals**.

You can find more information about the elements in Appendix C.

Period	The first element	The last element	Number of elements
1st period	$_1H$	$_2He$	2
2nd period	$_3Li$	$_{10}Ne$	8
3rd period	$_{11}Na$	$_{18}Ar$	8
4th period	$_{19}K$	$_{36}Kr$	18
5th period	$_{37}Rb$	$_{54}Xe$	18
6th period	$_{55}Cs$	$_{86}Rn$	32
7th period	$_{87}Fr$		

The first element, the last element and the number of elements in each period in the periodic table.

1.1. THE PERIODIC TABLE AND ELECTRON CONFIGURATION

The physical and chemical properties of the elements are directly related to their electron configurations. For example, chemical properties such as gaining, giving and sharing of electrons are dependent on the valence electrons and nucleus structure. As a result, chemical behaviors of the elements are closely related to the nucleus structure and electron configuration of the element. Elements in the same period contain different numbers of electrons in the valence shells.

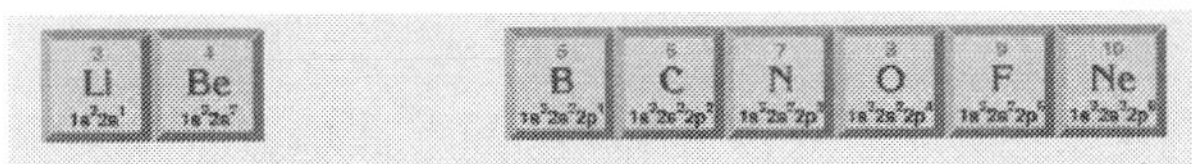

Electron configurations of the elements in the second period

For this reason, elements in the same period have different physical and chemical properties.

The valence electron configurations of the elements in the same group are the same. Therefore, elements in the same group show similar chemical behaviors in a chemical reaction, but their physical properties may gradually change.

The electron configurations of the elements in group 1A end with s^1 and the elements in group 2A end with s^2.

READING — Dmitri Ivanovich Mendeleev (1834-1907)

Mendeleev, the youngest of a family of seventeen, was born in Tobolsk, Siberia. His father was the director of a high school and his grandfather was the assistant of the first journalist in Siberia. After his father died, his mother moved to St. Petersburg to give him the best possible education.

Dmitri proved his worth in St. Petersburg when he prepared a thesis "The Combination of Alcohol and Water" (1856). Mendeleev met and worked with Bunsen and a lot of Western scientists, and participated in the Karlsruhe conference in Germany (1858). At this conference, there was intensive discussion about Avogadro's hypothesis. Dimitri then visited Pennsylvania to see the first oil wells. After returning to Russia, he developed a new commercial distillation system when he was 32 years old.

Mendeleev was appointed professor of inorganic chemistry at St. Petersburg University. His most important study was the periodic table that he developed using the regularities of chemical and physical properties of the elements.

In this study, Mendeleev estimated the existence of some elements which had not yet been discovered. The discovery, a few years later, of elements he had estimated made him a world famous chemist in the field of the periodic table.

His other studies, collected in 25 books, are very interesting as well. He organized knowledge on isomorphism, a study which supported the development of geochemistry.

Furthermore, he found the critical point and developed the "hydrate theory", which made him a great physical chemist.

Mendeleev was a member of about 70 academic and scientific committees. He regarded, his first responsibility as research and his second responsibility as learning. He worked as a teacher in most of the schools in St. Petersburg.

When Mendeleev published his periodic table for the first time, there were 63 elements. After his death, the number of elements had increased to 86. This quick increase was the result of the periodic table, the most important systemization of chemistry.

Although Mendeleev did not discover any new elements, the element with the atomic number 101 discovered by a committee of American scientists led by G.T Seaborg in 1955, was named mendelevium (Md) in honor of Dmitri Mendeleev.

1.2. PERIODS

It has already been mentioned that the horizontal rows are called **periods**. There are seven periods of which three are short, two are medium and two are long. Now, let us briefly examine these periods.

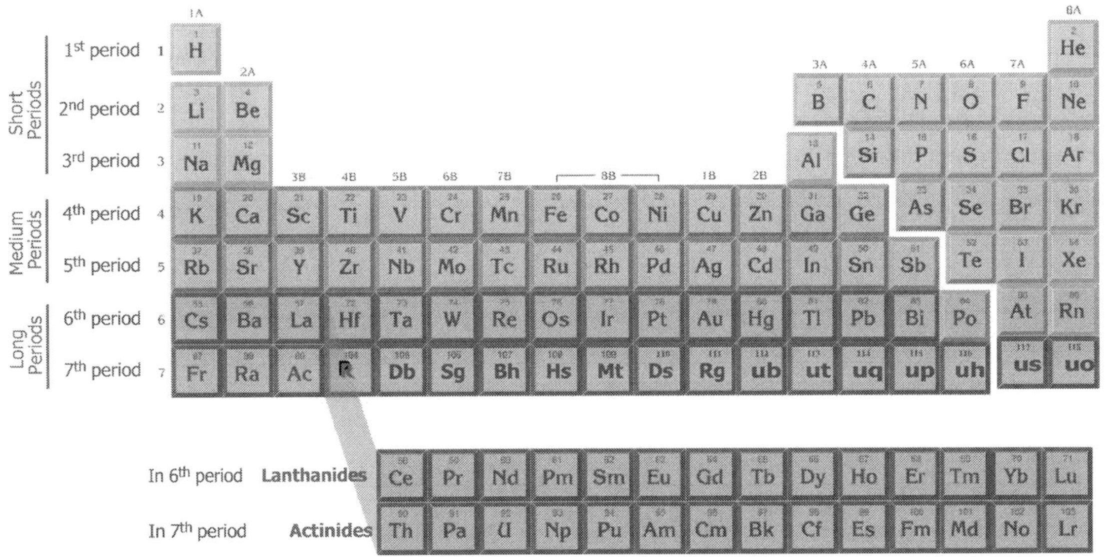

The periods of the periodic table.

The Short Periods

The first, second and third periods are short periods. There are two elements in the first period and eight elements in both the second and the third periods. The elements in the short periods constitute almost 97% of the earth's crust, oceans and atmosphere. Among these elements, helium (He) and neon (Ne) gases occur only in trace amounts in the atmosphere but argon (Ar) makes up about 1% of the atmsophere. The other elements are abundant.

The First Period

The first period is the smallest period and contains only two elements (hydrogen and helium). In the ground state of these elements, there is only one s orbital. Hydrogen occurs in a diatomic structure H_2, and helium (He) occurs in monoatomic structure. Both of them are in a gaseous state at room temperature. Since helium is chemically inert, it is placed in the noble gas family (even though its electron configuration ends with $1s^2$). Although hydrogen is placed in the group 1A, it shows nonmetallic behavior too.

The Second Period

There are eight elements in this period. Nitrogen and oxygen and fluorine occur as diatomic molecules in the atmosphere, such as N_2, O_2, and F_2. The gas neon exists in a monoatomic structure. After helium and hydrogen, Neon, has the lowest freezing point. Carbon (C) is basic element of the organic chemistry.

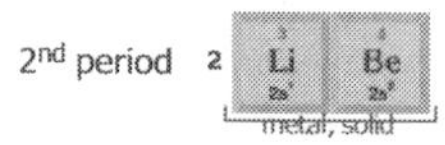

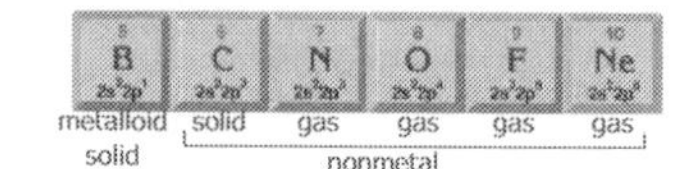

The Third Period

The elements in the third period also have one s and three p orbitals in their valence shells.

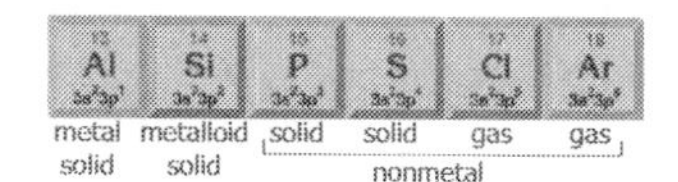

The Medium Periods

The fourth and fifth periods are the medium periods. Each of these periods contains 18 elements.

The Fourth Period

In this period, the valence electrons of the elements may be in 4s, 3d or 4p orbitals. This period starts with potassium (K) and ends with krypton (Kr).

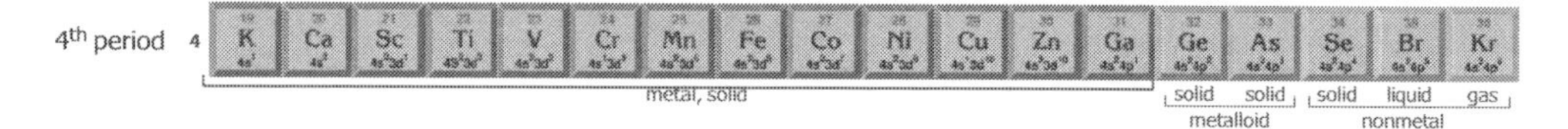

The Fifth Period

This period is completed by filling the 5s, 5p or 4d orbitals.

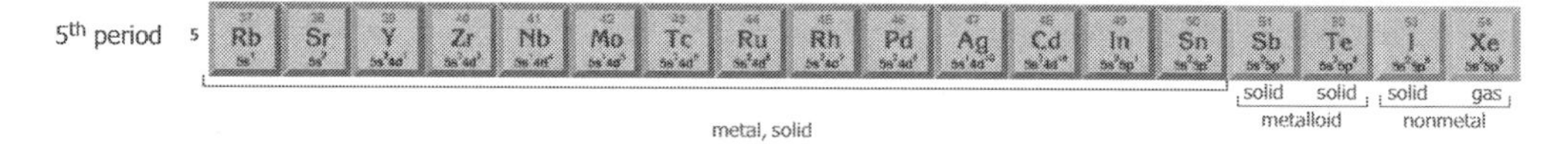

The Long Periods

The sixth and the seventh periods are the long periods. The sixth period includes 32 elements, and only 28 elements of the seventh period elements are known.

The Sixth Period

The sixth period starts by filling of 6s orbital with cesium (Cs) and barium (Ba).The 4f orbitals start to be occupied after one electron goes to the 5d orbitals.

The 14 elements starting with $_{57}$La, are called **lanthanides.** In this period, there are a total of 32 elements.

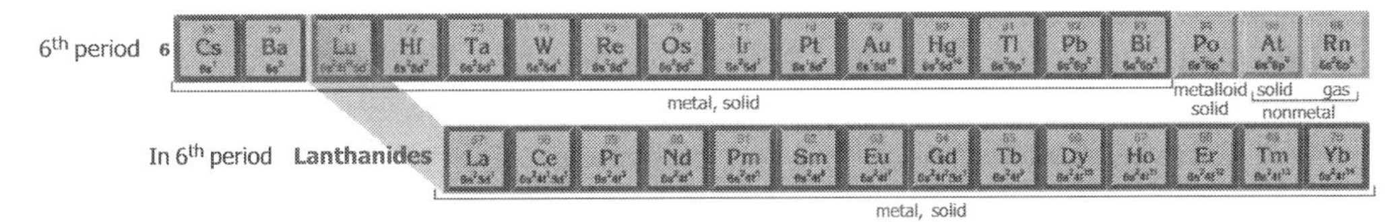

The Seventh Period

Currently, 28 elements filling the 7s, 6d and 5f orbitals have been discovered. In the seventh period, the 14 elements coming after $_{89}$Ac are called **actinides**. All the elements in the 6d block have been discovered, but not all of the elements in the 7p block.

Lanthanides and actinides are known as rare earth metals or inner transition metals.

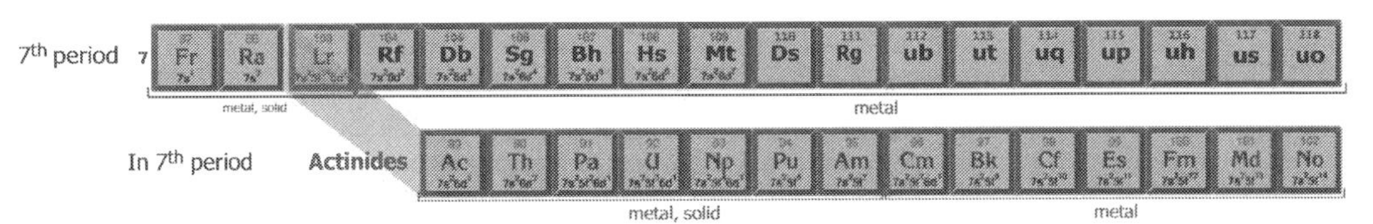

1.3. GROUPS

The periodic table contains the elements of the groups A and groups B. Among the elements of the groups A, group 1A and 2A elements belong to the s–block and group 3A, 4A, 5A, 6A, 7A and 8A elements belong to the p–block. The elements of the groups B belong to the d–block (transition metals) and f–block (inner transition metals).

In the periodic table, each A group has a special name.

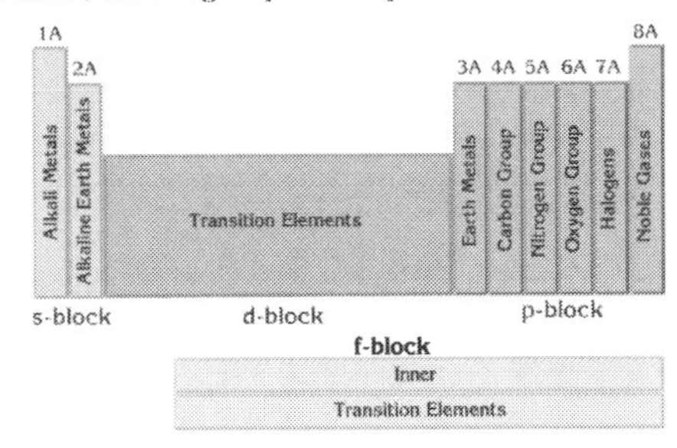

Families and the s, p, d, f blocks in the periodic table

Now let's examine the properties of some groups in the periodic table.

Group 1A : Alkali Metals (ns^1)

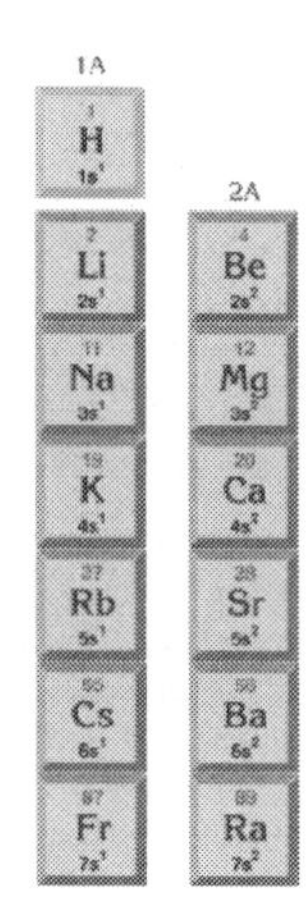

Alkali metals (group 1A) and alkaline earth metals (group 2A)

This group contains the elements hydrogen (H), lithium (Li), sodium (Na), potassium (K), rubidium (Rb), cesium (Cs), and francium (Fr). Although hydrogen is a non-metal, it is placed in this group because of its electron configuration, $1s^1$.

Alkali metals are chemically the most active metals. They have only one electron in their valence orbitals and their electron configuration ends with ns^1. They become positive one charged ions easily by giving off this valence electron in chemical reactions.

In nature, the hydrogen molecule occurs in a diatomic structure (H_2). It is in a gaseous state at room temperature. Hydrogen, whose characteristics are predominantly non-metallic, has a negative one charge in some compounds. All other members of this group, having the typical metallic characteristics, are solid at room temperature. Francium is only radioactive element in this group.

Group 2A : Alkaline Earth Metals (ns^2)

This group contains the elements beryllium (Be), magnesium (Mg), calcium (Ca), strontium (Sr), barium (Ba) and radium (Ra). After the alkali metals, they are the second most active metals. Their electron configurations end with ns^2. They become positive two charged ions by giving of their two valence electrons in chemical reactions. At room temperature, they occur in a monoatomic structure and they are solid at room temperature. Radium, a solid element, is the only radioactive member of this group.

Group 7A: Halogens (ns^2 np^5)

Halogens (group 7A) and noble gases (group 8A)

This group contains the elements fluorine (F), chlorine (Cl), bromine (Br), iodine (I) and astatine (At). These elements occur naturally in a diatomic structure (F_2, Cl_2, Br_2, I_2, At_2). Fluorine and chlorine are gases, bromine is a liquid and iodine is a solid. Astatine is a radioactive and solid element.

Halogens are the most active nonmetals. Therefore, they become negative one charged ions by gaining an electron to complete their valence shell in the chemical reactions.

Group 8A: Noble Gases (ns^2np^6)

This group contains the elements helium (He), neon (Ne), argon (Ar), krypton (Kr), xenon (Xe) and radon (Rn). Their electron configurations end with ns^2np^6, except helium, which ends with ns^2. Since their valence shells are full, they have no chemical activity and they are called **noble gases** or **inert gases**. They are the most stable elements in nature. Therefore, all other elements try to make their electron configurations similar to noble gases; either by gaining or losing electrons electrons. They occur in a monoatomic structure as colorless gases with low freezing points at room temperature.

1.4. FINDING GROUP AND PERIOD NUMBERS

We know that the elements are arranged in the order of increasing atomic numbers. Elements which have the same electron configuration in their valence shells are placed in the same group.

Knowing this, the place of any element in the periodic table can be easily found when its atomic number or the number of the electrons of the neutral atom or the electron configuration in ground state is known.

$1s^2, 2s^2, 2p^6, 3s^2, 3p^6, 4s^2, 3d^{10}, 4p^6, 5s^2, 4d^{10}, 5p^6, 6s^2, 4f^{14}, 5d^{10}, 6p^6, 7s^2, 5f^{14}, 6d^{10}, 7p^6$ ....

You can find electron configurations of all elements in Appendix A.

In order to find the group and period number of any atom, the ground state electron configuration is written first. Then the following points are taken into consideration.

1. The biggest principal quantum number (shell number) of any atom shows the period number of that atom.
2. If the electron configuration of an atom ends with s or p orbitals, this atom belongs to group A and the group number is determined by adding the number of electrons in the valence shell.

   If the electron configuration of an atom ends with the d orbital, this atom belongs to group B and the group number is determined by adding the electrons in the d orbital and electrons in the last s orbital.

   This rule doesn't work for 1B, 2B and 8B groups (there are three columns in group 8B). It is shown in table below

   All the elements of which the electron configurations end with the f orbital belong to group 3B.

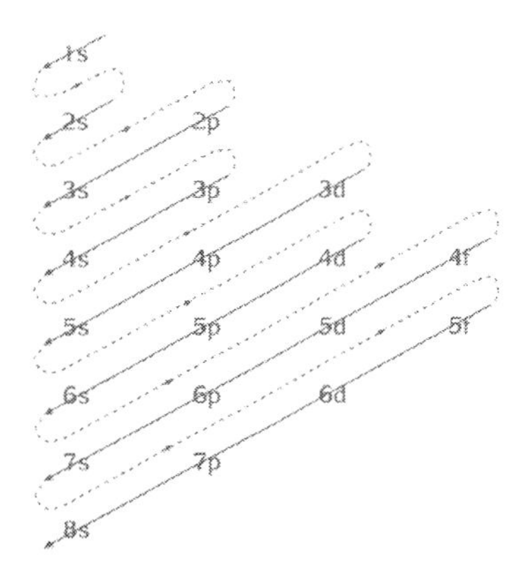

The Aufbau order of filling the atomic orbitals.

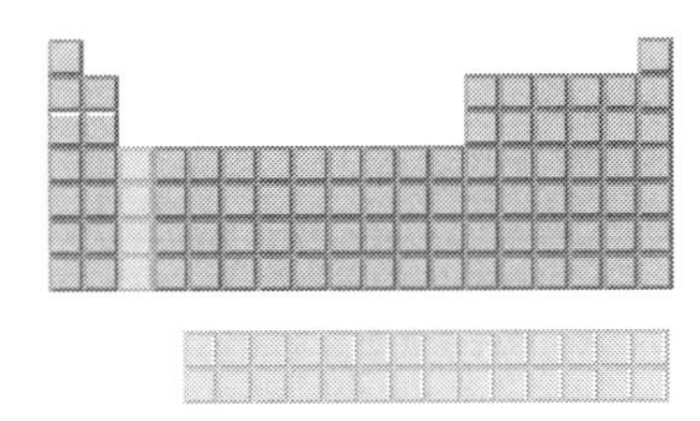

The group 3B includes 32 elements of which most are artificial.

According to the number of valence electrons, groups can be found as follows.

The principal quantum number		Group
n	n	
s^1	p^0	1A
s^2	p^0	2A
s^2	p^1	3A
s^2	p^2	4A
s^2	p^3	5A
s^2	p^4	6A
s^2	p^5	7A
s^2	p^6	8A

n is the biggest principal quantum number.

The principal quantum number		Group
n	n−1	
s^2	d^1	3B
s^2	d^2	4B
s^2	d^3	5B
s^1	d^5	6B
s^2	d^5	7B
s^2	d^6	8B
s^2	d^7	8B
s^2	d^8	8B
s^1	d^{10}	1B
s^2	d^{10}	2B

n is the biggest principal quantum number.

In th: light of this information, let us ε xamine t e examples given below.

Example 1

Determine the group and period numbers of:

a. Hydrogen, of which the electron configuration is $1s^1$

b. Helium, of which the electron configuration is $1s^2$

Solution

a.

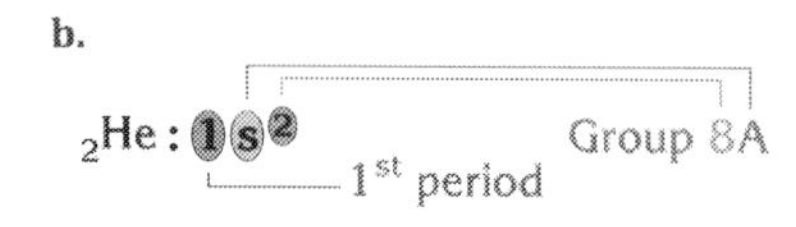

Since hydrogen has one valence electron, its group number is 1 and since the electron is in the s orbital, hydrogen is in group A. Therefore, H is in the 1A group. At the same time, because the principal quantum number (valence shell) of the s orbital is 1, H is in the 1st period. So, hydrogen is the element of 1st period and group 1A.

b.

$_2He : 1s^2$ Group 8A, 1st period

Helium is also in the 1st period because its valence electrons are in the 1st shell. It is expected that He element is in group 2A because its valence electrons are in s orbitals. However, the elements of which the valence shells are full are chemically inert and are called noble gases. The other noble gases constitute group 8A because of the electron configuration of ns^2np^6. Although the He element has the electron configuration of $1s^2$, it is in the 8A group. Thus, helium is a noble gas.

Example 2

Determine the places of aluminum element in the periodic table, if its electron configuration is $1s^22s^22p^63s^23p^1$

Solution

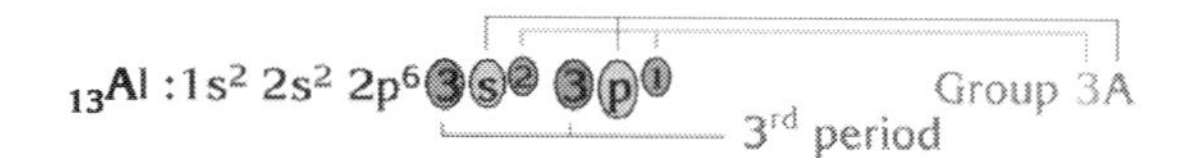

The biggest principal quantum number in the electron configuration of aluminum is three. Therefore, aluminum is in the 3rd period. The electron configuration of electrons in valence shell is $3s^23p^1$ and the number of valence electrons is three. The number of valence electrons determines the group number and having valence electrons in the s and p orbitals shows that the element is in the A groups. Thus, aluminum is in the 3A group. As a result, aluminum is in the 3rd period and group 3A of the periodic table.

Example 3

Determine the places of the elements $_{21}Sc$, $_{26}Fe$, $_{28}Ni$ and $_{29}Cu$ in the periodic table.

Solution

Electron configurations of transition elements end with ns (n-1)d. Its group number is equal to the total number of electrons in those shells. But this rule does not work for the elements with more than 8 electrons in those shells. In such cases, if the total number of electrons is equal to 9 or 10, the element is in 8B. If it is equal to 11, the element is in 1B and if it is equal to 12, the element is in the 2B group.

$ns^2(n-1)d^7 \rightarrow$ 8B group $ns^1(n-1)d^{10} \rightarrow$ 1B group

$ns^2(n-1)d^8 \rightarrow$ 8B group $ns^2(n-1)d^{10} \rightarrow$ 2B group

Given this, the placements of Sc, Fe, Ni and Cu in the periodic table can be found as follows.

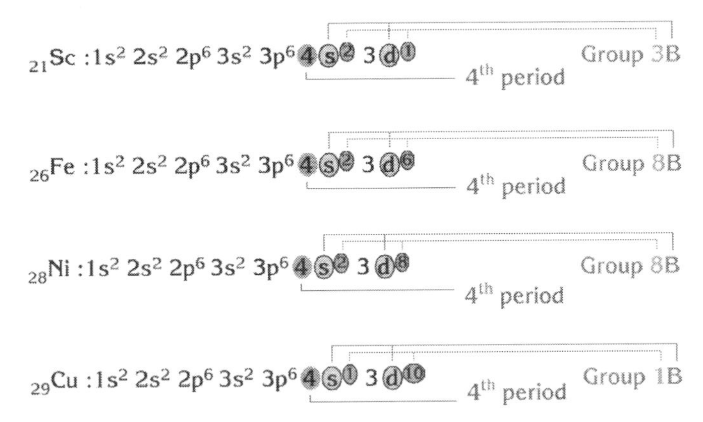

Example 4

What is the number of electrons of an element that is in 3rd period and group 4A of the periodic table?

Solution

The electron configuration of the element which is placed in the 3rd period and group 4A is $1s^22s^22p^63s^23p^2$. Therefore, the number of electrons is 14.

Exercise 1

The electron configuration of $_{58}Ce$ is $[Xe]6s^2\ 4f^2$. What are the group and period numbers of $_{58}Ce$?

Answer : *Group 3B and 6th period.*

Exercise 2

An X^{3+} ion has 10 electrons. What are the group and period numbers of the X atom of the periodic table?

Answer : *Group 3A and 3th period.*

Exercise 3

The atomic mass number of X is 70 and the number of neutrons of X is 39. What are the group and period numbers of element X in the periodic table?

Answer : *Group 3A and 4th period.*

2. THE PERIODIC TRENDS

An atom gains many physical and chemical properties as a result of the interactions between the number of protons and the valence electrons.

A strong or a weak attraction of the electrons by the atom effects several properties, such as atomic radius, density, melting and boiling points, and electron gain or loss ability.

Here, let us examine some properties. These are electronegativity, metallic and nonmetallic properties, atomic and ionic radius, ionization energy, acidity and basicity.

2.1 ELECTRONEGATIVITY

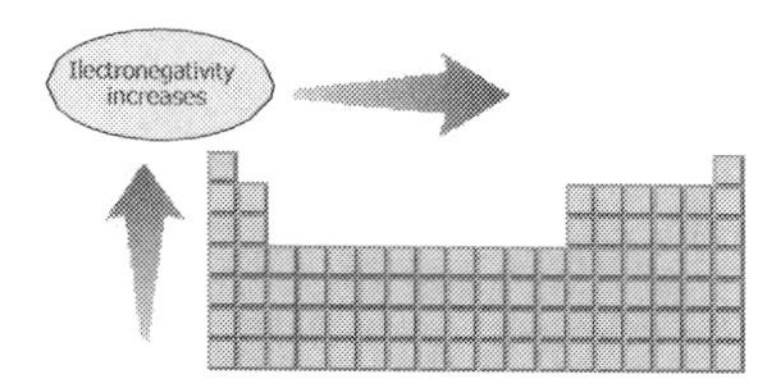

Electronegativity increases from left to right and from bottom to top.

Electronegativity is a measure of an atom's ability to attract electrons from a **covalent bond** in its molecule. The electronegativity of an atom depends on the charge of the nucleus and the distance between the nuclei and the electrons of a covalent bond. Therefore, electronegativity is closely related to ionization energy which expresses the ability of an atom to attract or loose electrons respectively.

Electronegativity is a relative quantity and it does not have a unit. Today, the most commonly used electronegativity scale today is Linus Pauling's scale, which is based on the values of bond energies.

According to this scale, the most active metal, francium, has a 0.7 value and the most active nonmetal fluorine has a 4.0 value. The electronegativity value of the other elements are between 0.7 and 4.0. The electronegativities of the elements are shown in the figure below.

The electronegativity of an atom depends on the radius of the atom. The atomic radius decreases and attraction exerted on valence electrons by the nucleus increases from left to right in a period. Atomic radius increases and attraction exerted on valence electrons by nucleus decreases from top to bottom. Therefore, electronegativity increases from left to right and decreases from top to bottom in the periodic table.

Bond	Electronegativity difference	Bond energy (kJ/mol)
H – F	1.80	568.5
H – Cl	0.80	430.5
H – Br	0.62	367.9
H – I	0.28	301.0

The electronegativity differences of bonds which are formed between hydrogen and halogens are shown. As seen, when the electronegativity differences increase, the bonds become stronger.

In the compounds of hydrogen with halogens, an increase in the atomic radius of the halogens causes a decrease in the electronegativity which also causes the strengths of the bonds to become weaker.

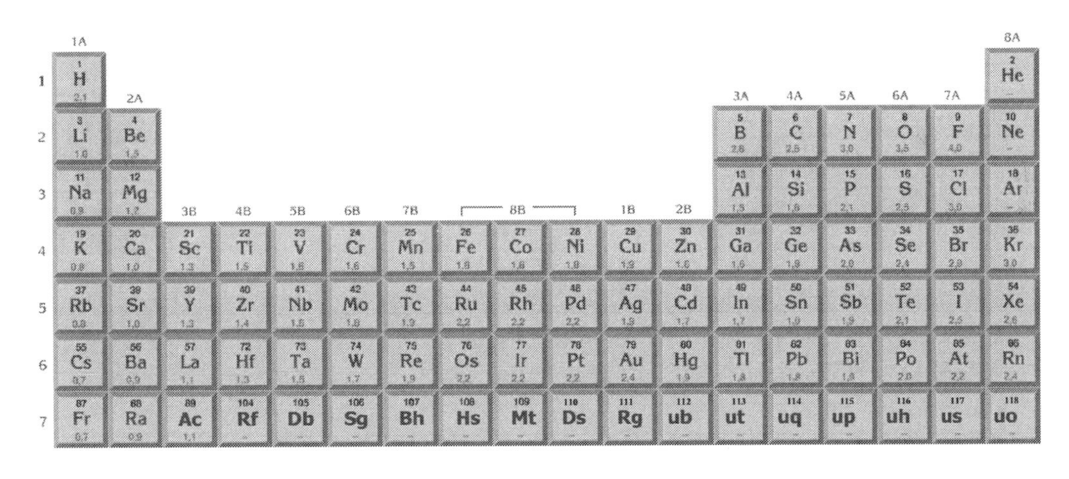

The electronegativity values of the elements in the periodic table.

2.2. METALLIC AND NONMETALLIC PROPERTIES

When we classify the elements as metals and nonmetals we see that metals occupy very big part (about 80%) of the periodic table. The elements in B groups (transition elements, actinides and lanthanides) and the elements in the groups, 1A, 2A and 3A (except hydrogen and boron) are **metals**. Only the eleven elements H, C, N, O, P, S, Se, F, Cl, Br and I are **nonmetals** and the elements in group 8A are **noble gases**. However, among these elements, B, Si, Ge, As, Sb, Te, Po and At are **metalloids** and Sn, Pb and Bi and Be have metallic properties.

At room temperature, all metals have a silvery–luster and are in the solid state (except Hg which is in the liquid state). Nonmetals, which are dull, can be found in the solid state such as S and I_2, in the liquid state (Br_2) and in the gaseous state, like N_2, O_2, F_2 and Cl_2.

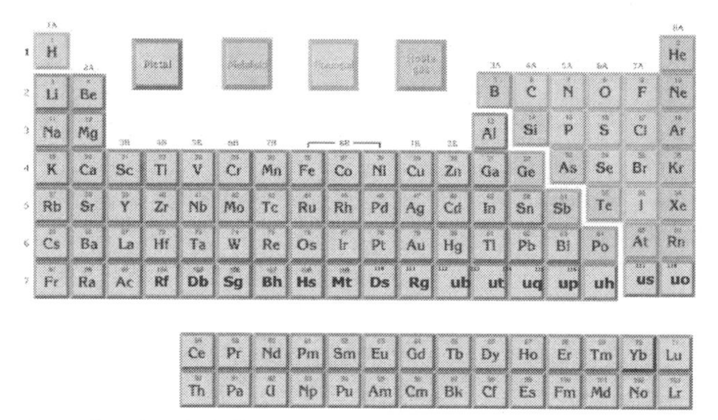

Metals are placed on the left side of the periodic table and non metals are placed on the right. Metalloids are placed between metals and nonmetals.

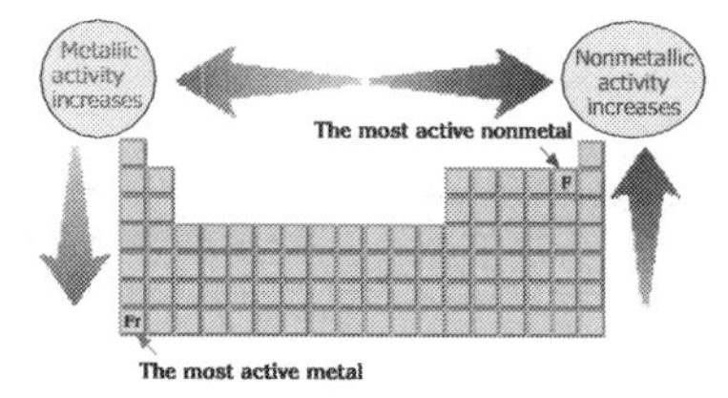

The metallic and nonmetallic properties in periodic table.

Metals form alloys with each other. They form ionic compounds with nonmetals. Nonmetals form only covalently bonded compounds with each other. As a result, metals become only positively (+) charged ions, whereas nonmetals become either negatively (–) or positively (+) charged ions.

Additionally, metals conduct heat and electricity well, whereas nonmetals do not. The metallic and nonmetallic activities of elements are closely related to the electronegativities. More electronegative nonmetals are more active and vice versa.

Metallic and nonmetallic properties are related to the number of valence electrons and the radius of an atom. Within a period, as the metallic properties decrease from left to right, the nonmetallic properties increase. Within a group as the metallic properties increase, the nonmetallic properties decrease from top to bottom. If the above trends are considered, francium, Fr, would be expected to have most metallic properties. However, since Fr is a radioactive element, not all of its properties have been determined yet.

On the other hand, the element with the most nonmetallic properties is fluorine, F. Moreover, the metalloids show either metallic or nonmetallic properties, depending on the conditions.

2.3. ATOMIC RADIUS

The physical properties of the elements, such as melting point, boiling point and density are related to the atomic radius of the elements. Also, the atomic radius directly affects the ability of an atom to gain and lose electrons. The atomic radius is practically defined by assuming the shape of the atom as a sphere. **The atomic radius** is the distance between the nucleus and the outermost electron. But it is impossible to measure the atomic radius by separating the atoms from each other.

The atomic radius of a metal is equal to half the distance (r_1) between the nuclei of the neighbouring metal atoms.

1 meter is equal to 10^{12} picometers.

Atomic Radius within a Group

Since the number of shells increases in the same group from top to bottom (by the period number increases), the atomic radius also increases. This means that the electron cloud around the nucleus becomes larger. The increase in the number of electrons causes them occupy a new energy level and orbitals. A higher energy level is always further from nucleus. Within a period, if the number of protons and electrons increases, the nuclear attraction force increases. This attraction force prevents an enormous increase in atomic radius.

The radii of atoms of group 2A elements increase when their atomic numbers increase, as in other groups.

Atomic Radius Within a Period

The atomic radius usually decreases from left to right in a period. It may be thought that the atomic radius has to increase because of the increase in the number of electrons in a period (in the same shell). However, the number of protons also increases by as much as that of electrons. Increasing the number of protons increases the nuclear attraction force on the electrons. Thus, since the intensity of nuclear attraction force per one electron increases, the atomic radius decreases from left to right in a period.

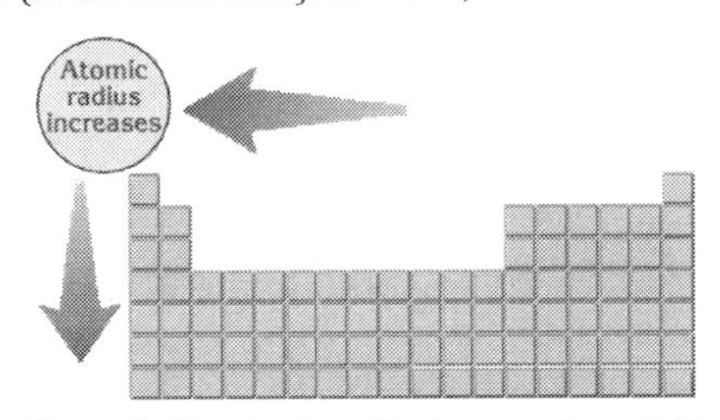

Generally the atomic radius increaes from right to left and from top to the bottom.

Changes in the radii of atoms of the second period elements according to their atomic numbers.

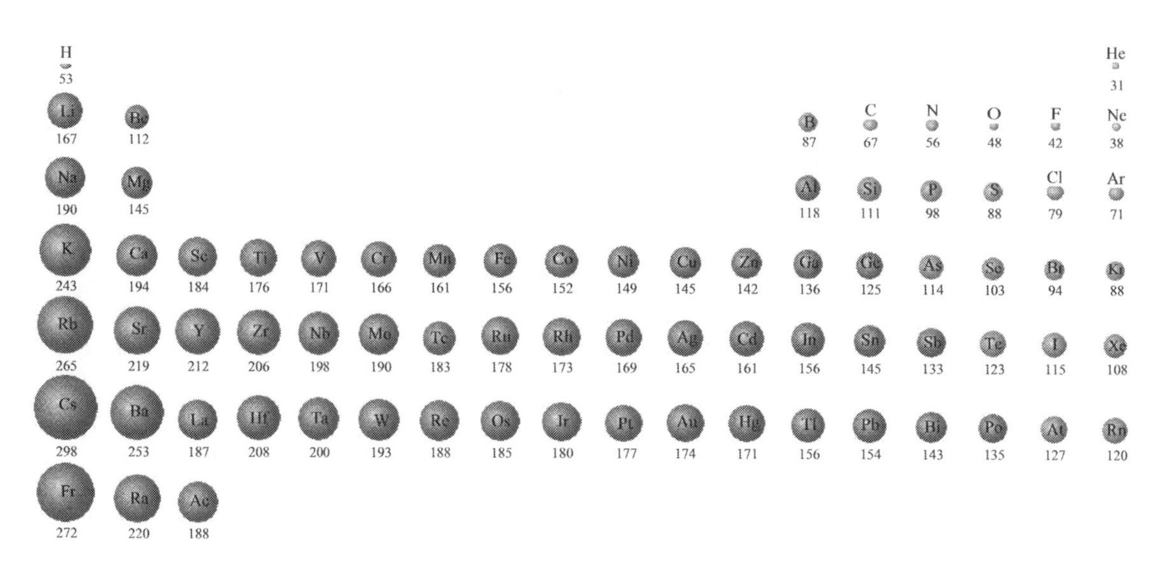

Spherical illustrations of the atoms of the elements. The calculated values of atomic radii are given in picometers.

The changes in the atomic radius of transition elements are not as dramatic as in A group elements. This is because the d orbitals are filled after filling the s orbitals which are at a higher energy level (being further from the nucleus). For example, after filling the $4s^2$ orbital, the 3d orbitals at a lower energy level are occupied. So, there is no big change in the atomic radii of consecutive transition elements in the 4th period.

2.4. IONIC RADIUS

The radii of the positive ions, cations, are smaller than those of their parent neutral atoms. A decrease in the number of outer electrons means that the rest of the electrons are attracted more strongly by the protons. In other words, since the attraction force per electron increases, the radius becomes smaller. For example, the radius of the neutral Na atoms is 190 pm and the ionic radius of the cation Na^+ is 95 pm. When the neutral Na atom becomes a Na^+ ion by giving off one electron, the radius approximately halves.

The radii of the negative ions, anions, are bigger than those of their parent neutral atoms. The addition of an electron or electrons in the formation of an anion increases the repulsive forces between the outer electrons. Therefore, the radii of anions are bigger than those of their parent nonmetal atoms. For example, the ionic radius of F^- is 136 pm, which is about two times greater than the radius of neutral F, 64 pm.

The atomic radii of isoelectronic ions and their neutral atoms are not equal because their nuclei have different numbers of protons.

Since more protons means more attraction, more protons in the nuclei of isoelectronic ions or atoms have a smaller atomic radius.

In chemical reactions, metals such as lithium forms positively charged ions by giving off electrons and their atomic radii becomes smaller. On the other hand, nonmetals such as fluorine form negatively charged ions by gaining electrons and their atomic radii become bigger.

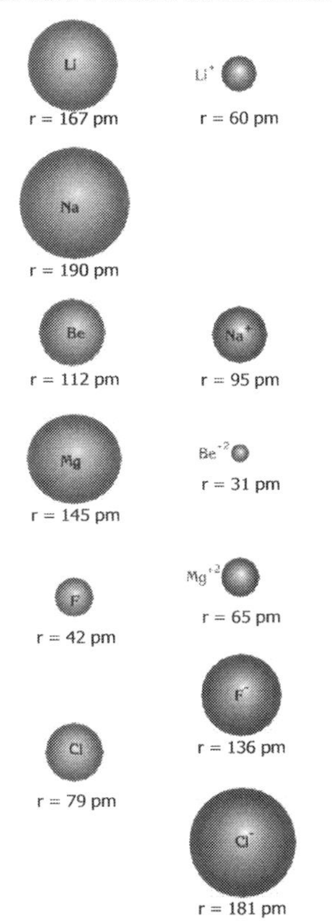

The radii of cations are smaller than those of their parent atoms and the radii of anions are bigger than those of their parent atoms.

	N^{3-}	O^{2-}	F^-	Na^+	Mg^{2+}	Al^{3+}
Radius of ions (pm):	171	140	136	95	65	50
Electron number :	10	10	10	10	10	10
Atomic number :	7	8	9	10	11	12

The atomic radii of isoelectronic ions are given here. As is easily seen, they radii become smaller as their atomic numbers (the charges of the nuclei) increase.

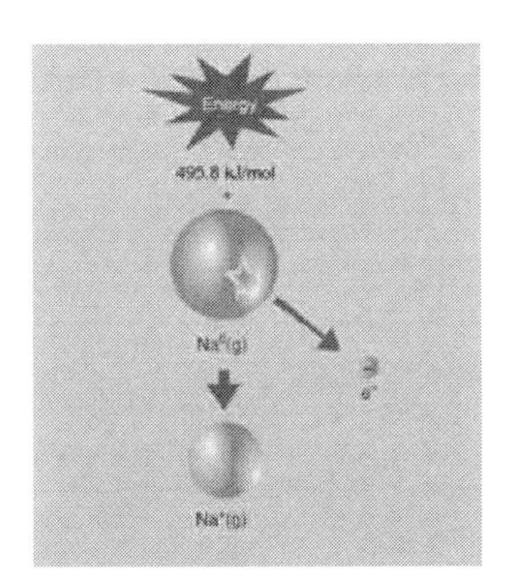

The first ionization energy is equal to 495.8 kJ/mol.

2.5. IONIZATION ENERGY

Ionization energy is the energy required to remove an electron from an atom in its ground state in the gas phase. This energy shows the degree of attractive force that the nucleus has on the electron. In order to accomplish ionization, the atom cannot be in neither a solid state nor a liquid state. The removable electron is the most loosely held electron.

The amount of energy required to remove one electron from the valence shell of a neutral atom in a gaseous state is defined as **the first ionization energy**, and denoted by I_1. The first ionization energies, I_1, and corresponding equations for the elements are given below.

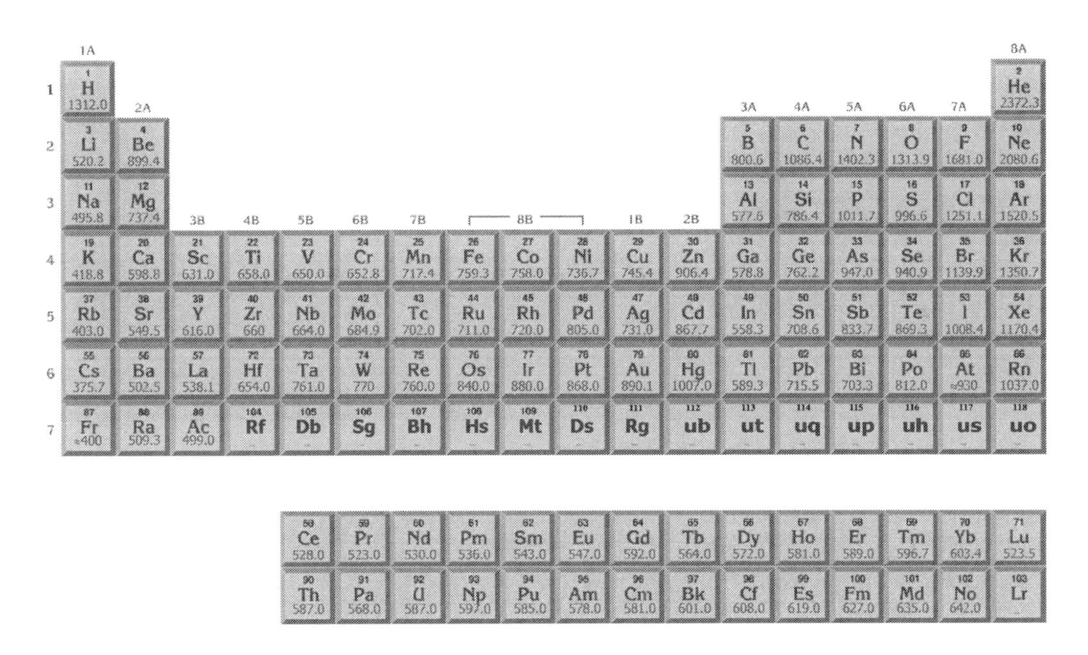

	1A	2A	3B	4B	5B	6B	7B	8B	8B	8B	1B	2B	3A	4A	5A	6A	7A	8A
1	1 H 1312.0																	2 He 2372.3
2	3 Li 520.2	4 Be 899.4											5 B 800.6	6 C 1086.4	7 N 1402.3	8 O 1313.9	9 F 1681.0	10 Ne 2080.6
3	11 Na 495.8	12 Mg 737.4											13 Al 577.6	14 Si 786.4	15 P 1011.7	16 S 996.6	17 Cl 1251.1	18 Ar 1520.5
4	19 K 418.8	20 Ca 598.8	21 Sc 631.0	22 Ti 658.0	23 V 650.0	24 Cr 652.8	25 Mn 717.4	26 Fe 759.3	27 Co 758.0	28 Ni 736.7	29 Cu 745.4	30 Zn 906.4	31 Ga 578.8	32 Ge 762.2	33 As 947.0	34 Se 940.9	35 Br 1139.9	36 Kr 1350.7
5	37 Rb 403.0	38 Sr 549.5	39 Y 616.0	40 Zr 660	41 Nb 664.0	42 Mo 684.9	43 Tc 702.0	44 Ru 711.0	45 Rh 720.0	46 Pd 805.0	47 Ag 731.0	48 Cd 867.7	49 In 558.3	50 Sn 708.6	51 Sb 833.7	52 Te 869.3	53 I 1008.4	54 Xe 1170.4
6	55 Cs 375.7	56 Ba 502.5	57 La 538.1	72 Hf 654.0	73 Ta 761.0	74 W 770	75 Re 760.0	76 Os 840.0	77 Ir 880.0	78 Pt 868.0	79 Au 890.1	80 Hg 1007.0	81 Tl 589.3	82 Pb 715.5	83 Bi 703.3	84 Po 812.0	85 At ≈930	86 Rn 1037.0
7	87 Fr ≈400	88 Ra 509.3	89 Ac 499.0	104 Rf –	105 Db –	106 Sg –	107 Bh –	108 Hs –	109 Mt –	110 Ds –	111 Rg –	112 ub –	113 ut –	114 uq –	115 up –	116 uh –	117 us –	118 uo –

58 Ce 528.0	59 Pr 523.0	60 Nd 530.0	61 Pm 536.0	62 Sm 543.0	63 Eu 547.0	64 Gd 592.0	65 Tb 564.0	66 Dy 572.0	67 Ho 581.0	68 Er 589.0	69 Tm 596.7	70 Yb 603.4	71 Lu 523.5
90 Th 587.0	91 Pa 568.0	92 U 587.0	93 Np 597.0	94 Pu 585.0	95 Am 578.0	96 Cm 581.0	97 Bk 601.0	98 Cf 608.0	99 Es 619.0	100 Fm 627.0	101 Md 635.0	102 No 642.0	103 Lr

First ionization energies of the elements in kJ/mol.

$Na(g) + 495\ kJ \longrightarrow Na^+(g) + e^-$ $\quad I_1 = 495.8\ kJ/mol$

$Mg(g) + 737.7\ kJ \longrightarrow Ca^+(g) + e^-$ $\quad I_1 = 737.7\ kJ/mol$

$F(g) + 1681.0\ kJ \longrightarrow F^+(g) + e^-$ $\quad I_1 = 1681.0\ kJ/mol$

$Ne(g) + 2080.6\ kJ \longrightarrow Ne^+(g) + e^-$ $\quad I_1 = 2080.6\ kJ/mol$

The energy required to remove the second electron and the third electron are called **the second ionization energy (I_2)** and **the third ionization energy (I_3)**, respectively.

$X(g) + Energy \longrightarrow X^+(g) + e^-$ (1st ionization energy)

$X^+(g) + Energy \longrightarrow X^{2+}(g) + e^-$ (2nd ionization energy)

$X^{2+}(g) + Energy \longrightarrow X^{3+}(g) + e^-$ (3rd ionization energy)

During the removal of each electron, since the nuclear charge remains constant and the number of electrons decreases, the nuclear attraction force per electron increases. That means that the remaining electrons are attracted even more strongly, and as a result, the following relation is obtained for successive ionization energies.

$I_1 < I_2 < I_3 < I_4 .... < I_n$

In other words, each successive ionization requires more energy. As an example, the first, the second and the third ionization energies of magnesium, Mg, are given below.

$Mg(g) \longrightarrow Mg^+(g) + e^-$ $\quad I_1 = 737.4\ kJ/mol$

$Mg^+(g) \longrightarrow Mg^{2+}(g) + e^-$ $\quad I_2 = 1454.6\ kJ/mol$

$Mg^{2+}(g) \longrightarrow Mg^{3+}(g) + e^-$ $\quad I_3 = 7720.5\ kJ/mol$

If the numerical values of these ionization energies are taken into consideration, the value of the second ionization energy is about twice the value of the first one. However the third ionization energy is about 5.3 times greater than the second ionization energy.

When the magnesium atom, $_{12}Mg$ becomes Mg^{2+} by giving off two electrons, its electron configuration will be similar to the electron configuration of Neon, Ne. The Mg^{2+} ion has noble gas stability. Since the removal of an electron from Mg^{2+} ion is very difficult, there is a big jump between the second and third ionization energies. Such a big jump is used to find out the number of valence electrons and the group number for the elements in the A groups.

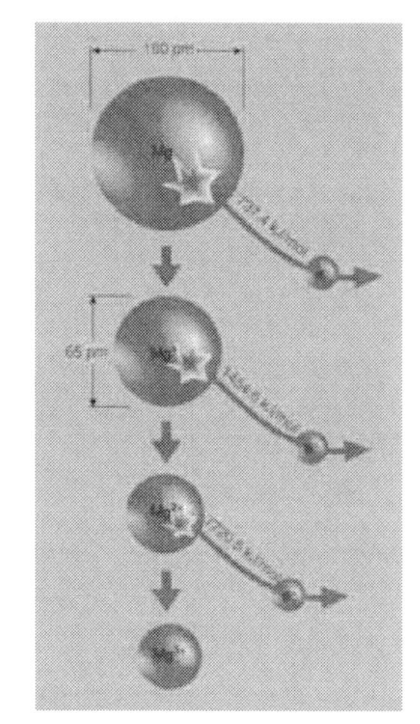

The third ionization energy of magnesium is very high because the electron configuration of Mg^{2+} has noble gas stability.

Example 5

In the table below, the first five ionization energies of Na, Mg and Al are given. Find the number of valence electrons and the group numbers of these elements.

Ionization Energy (kJ/mol)

	I_1	I_2	I_3	I_4	I_5
Na	495.8	4564	6905	9530	13401
Mg	737.4	1455	7720	10538	13601
Al	577.6	1814	2746	11566	14801

Solution

For the element sodium,Na, the first ionization energy is 495.8 kJ/mol and the second ionization energy is about 5.5 times greater than the first ionization energy. Thus, the sodium, Na, atom gets a noble gas electron configuration after giving off one electron.

It can be concluded that since the element sodium, Na, has only one valence electron, it is in group 1A.

Similar big jumps are seen for magnesium, Mg, between I_3 and I_2, and for aluminum, Al, between I_4 and I_3.

As a result the number of valence electrons of Mg and Al are two and three respectively. That is, magnesium, Mg, is in group 2A, aluminum, Al, is in group 3A.

Exercise 4

The first five ionization energies of the silicon atom are 786.4, 1580, 3231, 4347 and 16101 in kJ/mol respectively. Find the number of valence electrons and the group number of atom Si.

Answer : *4 valence electrons and group 4A.*

Atomic Number	Symbol	Ionization Energies (kJ / mol)									
		I_1	I_2	I_3	I_4	I_5	I_6	I_7	I_8	I_9	I_{10}
1	H	1312.0									
2	He	2372.3	5242								
3	Li	520.2	7290	11800							
4	Be	899.4	1756	14831	20979						
5	B	800.6	2424	3658	24988	32767					
6	C	1086.4	2349	4615	6212	37758	47192				
7	N	1402.3	2862	4573	7478	9434	53174	64247			
8	O	1313.9	3390	5313	7445	10973	13305	71211	83930		
9	F	1681.0	3373	6040	8398	11002	15136	17836	91876	105984	
10	Ne	2080.6	3959	6166	9355	12181	15215	19943	22931	114653	130069

The first ten ionization energies for the first ten elements of the periodic table.

The ionization energy is a characteristic property for an element.

There are some relationships between the ionization energies of the elements in the periodic table. Now let us examine these relationships briefly here.

1. Variation of the Ionization Energy within a Group

As the atomic radius increases from top to bottom in a group, the valence electrons become more further away from the nucleus and the nuclear attraction forces on these electrons decrease. Therefore, as the atomic radius increases, the amount of energy required to remove an electron decreases. As a result, we can say that within a group ionization energy of elements decrease from top to bottom.

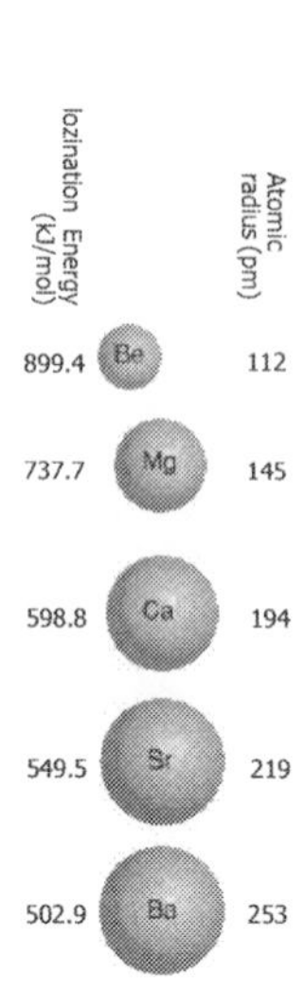

2. Variation of the Ionization Energy within a Period

In a period, since the atomic radius decreases from left to right, the ionization energy generally increases.

The elements of group 1A (alkali metals), includes elements which have the greatest atomic radii; therefore, the ionization energies of alkali metals are the lowest in every period. The elements of noble gases (group 8A) have the highest ionization energies, which means that the noble gases have a very stable electronic structure.

According to spherical symmetry (explained right), the variation of ionization energies in a period is as follows:

1A < 3A < 2A < 4A < 6A < 5A < 7A < 8A

In group B elements (transition and inner transition metals), there are no fixed jumps between the two consecutive ionization energies. That is why we cannot talk about the relationships of ionization energies of these elements.

Spherical Symmetry

The elements in group 2A have higher ionization energies than the elements in group 3A and the elements in group 5A have higher ionization energies than the elements in group 6A of the same period. Additionally, the elements in group 8A have the highest ionization energy in the same period. All of these can be explained by spherical symmetry. Spherical symmetry is a state in which the valence orbital or orbitals of an atom are filled or half-filled. Homogeneous distribution of electrons gives an unexpected stability to atom. That is why atoms in group 2A and 5A have higher ionization energy than expected. Atoms with spherical geometry have electronic configurations ending with ns^2, np^3, np^6 or ns^1, nd^5, nd^{10}, nf^7 and nf^{14}.

Example 6

Which atom of the $_3$Li and $_{19}$K would have the higher first ionization energy?

Solution

Let us first write the electron configurations of these atoms.

$_3Li : 1s^2\ 2s^1$ and $_{19}K : 1s^2\ 2s^2\ 2p^6\ 3s^2\ 3p^6\ 4s^1$

From the electron configurations we can see that lithium is in the 2nd period, group 1A, and potassium is in the 4th period, group 1A. Since lithium is above potassium in the same group, the atomic radius of lithium is smaller than that of potassium but the first ionization energy is higher.

Exercise 5

Which atoms in the $_{11}$Na- $_{16}$S and $_4$Be- $_5$B pairs would have the higher first ionization energy?

Answer : *S and Be.*

2.6. ACIDITY AND BASICITY

$Na_2O(s) + H_2O(l) \rightarrow 2NaOH(aq)$

$2NaOH(aq) \rightarrow 2Na^+(aq) + 2OH^-(aq)$

Basic oxides form OH^- ions when dissociated in water.

$Cl_2O(g) + H2O(l) \rightarrow 2HClO(aq)$

$2HClO(aq) \rightarrow 2H^+(aq) + 2ClO^-(aq)$

Acidic oxides form H^+ ions when dissociated in water.

The acidity and basicity of an oxide of an element depends on the electronegativity of that element. The greater the electronegativity the more acidic the oxide of the element, and the less the electronegativity, the more basic the oxide of the element.

Let us examine the oxides of the elements of the third period. These oxides, from 1A to 7A, are Na_2O, MgO, Al_2O_3, SiO_2, P_2O_3, SO_2 and Cl_2O. Na_2O and MgO are basic oxides, Al_2O_3 is an amphoteric oxide and SiO_2, P_2O_3, SO_2 and Cl_2O are acidic oxides. Thus, acidity of the elements increases and the basicity of the elements decreases from 1A through 7A in the third period. In general, the acidity of oxides of the elements increases from left to right in a period, and the basicity of the oxides of the elements decreases.

Since the electronegativity of elements decreases from top to bottom in a group, the basicity of the oxides of the elements increases and the acidity of the oxides of the elements decreases.

Acidic property increases

Basic property increases

1A	2A	3A	4A	5A	6A	7A
	BeO			N_2O_5		OF_2
	MgO			P_4O_{10}		Cl_2O_7
	CaO			As_2O_5		Br_2O_7
	SrO			Sb_2O_5		I_2O_7
	BaO			Bi_2O_5		At_2O_7

The change in the basic and acidic properties here shown for the oxides of main group elements which have the highest oxidation state number. The oxides in the blue colored regions are basic (metallic) oxides and the oxides in the red colored regions are acidic (nonmetallic) oxides. The oxides in both blue and red colored regions are amphoteric oxides (the oxides of amphoteric metals).

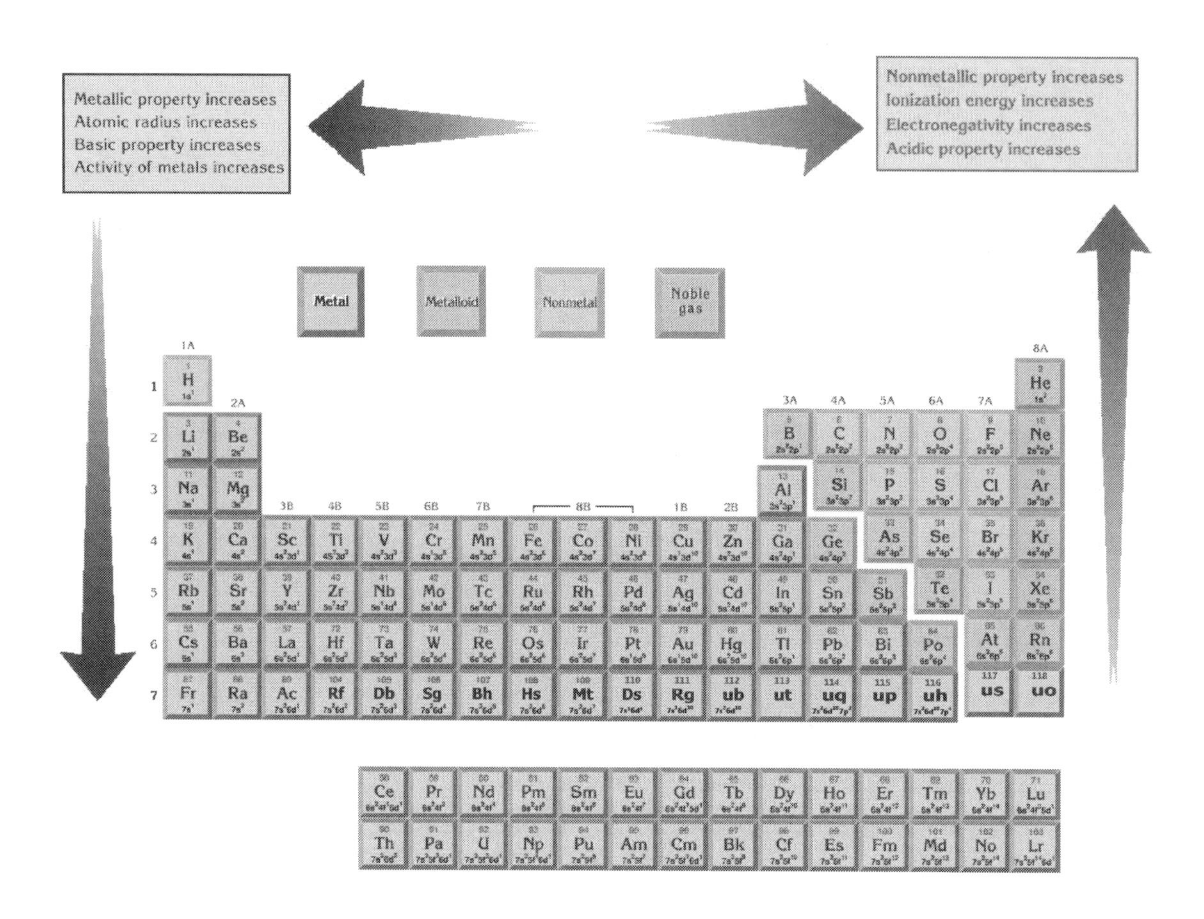

The general properties of elements in the periodic table

SUPPLEMENTARY QUESTIONS

1. The elements, calcium, strontium and barium are a triad observed by Döbereiner. Below there are some properties of this triad.

	melting point (°C)	density (g/mL)
$_{20}Ca$	a	1.54
$_{38}Sr$	771	b
$_{56}Ba$	717	3.60

What can you say about the values of a and b?

2. What is the law of octaves? Explain.
3. What were the periodic properties of the elements upon which Mendeleev and Meyer established their periodic table of the elements?
4. Mendeleev left some places empty in his periodic table, when he was arranging the elements according to increasing atomic weight. Why did he do so? Explain.
5. Why did Mendeleev place the element cobalt before the element nickel even though cobalt is heavier?
6. In the periodic table, as the atomic number increases, generally the atomic mass also increases. Which elements are exceptions to this generalization?
7. Which misunderstanding about the periodic table was eliminated by the Moseley's studies?
8. Define the terms, 'period', 'group', 'main group', and 'sub-group'.
9. "Each period starts with a metal". Is there any exception to this?
10. How many elements are there in first five periods respectively?
11. Which elements constitute the first period? Explain their properties.
12. Where are the transition and inner transition elements in the periodic table? The inner transition elements (lanthanides and actinides) are placed in a special region in the periodic table. Explain the reason for this.
13. What are the alkali metals?
14. The hydrogen element is placed in the 7A group in some periodic tables. Explain why this is so.
15. What are the elements with similar chemical properties to calcium?
16. Which elements are found in the gaseous state in the form of monoatomic structure in room conditions?
17. Find the group numbers of the elements with electron configurations ending with s^1, s^2, s^2p^5 and s^2p^6.
18. Although helium has an electron configuration of $1s^2$, it is shown in the group of noble gases. Explain.
19. Write down the ground state electron configurations of the following elements: $_5B$, $_{15}P$, $_{30}Zn$ and $_{31}Ga$.
20. Write down the ground state electron configuration of the element $_{47}Ag$.
21. Find the places of the elements for which the ground state electron configurations are given below.

    a. $1s^2$ b. $[He]2s^22p^5$ c. $[Ar]3d^34s^2$

    d. $[Ne]3s^23p^1$ e. $[Ar]3d^{10}4s^2$ f. $[Kr]4d^{10}5s^25p^5$

22. What is the atomic number of the element that has one electron in the 3rd energy level?
23. Which of the elements in the second period ($_3Li$, $_4Be$, $_5B$, $_6C$, $_7N$, $_8O$, $_9F$ and $_{10}Ne$) have half-filled s or p orbitals?
24. Find the numbers of unpaired electrons for the elements $_{13}Al$, $_{14}Si$, $_{15}P$ and $_{25}Mn$.
25. Find the atomic number of the element with a total of 13 electrons in p orbitals in its ground state.
26. Find the group number of the element that has 7 electrons in d orbitals in its ground state?
27. Find the total numbers of p electrons in the elements $_{14}Si$, $_{20}Ca$ and $_{33}As$.
28. For a neutral $^{80}_{35}Br$ atom:

    a. Find the number of neutrons and protons in the nucleus.

    b. How many s electrons does it have in total?

    c. How many electrons does it have in 4p orbitals?

    d. Which type of orbitals has the most electrons?

29. According to the information given below, arrange the atomic numbers of the X, Y, Z and T elements.

I. The element X has 9 electrons in s orbitals.

II. All of the orbitals of Y in 3rd energy level are full filled.

III. The element Z has 3 electrons in d orbitals.

IV. The element T has 10 electrons in p orbitals.

30. Write down the electron configuration of the following ions.

a. ${}_{7}N^{3-}$ **b.** ${}_{16}S^{2-}$ **c.** ${}_{26}Fe^{2+}$ **d.** ${}_{35}Br^{7+}$

31. Answer the questions below for these ions:

${}_{16}S^{2-}$, ${}_{20}Ca^{2+}$, ${}_{15}P^{3-}$, ${}_{19}K^{+}$, ${}_{17}Cl^{+}$, ${}_{25}Mn^{7+}$

a. Write down their electron configurations.

b. Which of these are isoelectronic with ${}_{18}Ar$?

32. The ions ${}_{21}X^{3+}$ and Y^{2-} have the same electron configuration. Find the place of Y in the periodic table.

33. How many electrons does ${}_{12}Mg^{2+}$ have in its s orbitals?

34. Find the numbers of valence electrons for the elements magnesium (${}_{12}Mg$), oxygen (${}_{8}O$), silicon (${}_{14}Si$) and zinc (${}_{30}Zn$).

35. Among the 4th period elements, which of them have 4 valence electrons?

36. State which of the following elements are metal, nonmetal or metalloids:

a. He **b.** K **c.** Mg **d.** Al

e. Ge **f.** Cl **g.** Sn **h.** Fe

37. Compare the metallic properties of the following elements; ${}_{19}K$, ${}_{11}Na$, ${}_{12}Mg$ and ${}_{6}C$.

38. Among the neutral atoms ${}_{7}N$, ${}_{8}O$, ${}_{11}Na$ and ${}_{12}Mg$, which one has the greatest atomic radius?

39. Among the elements (${}_{2}He$, ${}_{10}Ne$, ${}_{18}Ar$, ${}_{36}Kr$, ${}_{54}Xe$ and ${}_{86}Rn$) which one has the greatest atomic radius?

40. Are there any elements whose first ionization energy is smaller than its second ionization energy? Explain.

41. Among the elements, ${}_{20}Ca$, ${}_{15}P$ and ${}_{17}Cl$, which one has the lowest first ionization energy?

42. Compare the elements ${}_{4}Be$, ${}_{5}B$, ${}_{7}N$ and ${}_{8}O$ in terms of their first ionization energies.

43. Arrange the elements, ${}_{9}F$, ${}_{10}Ne$ and ${}_{11}Na$ in decreasing order of their first ionization energies.

44. The first ionization energy of Mg is 737.4 kJ/mol and the second ionization energy of Mg is 1454.6 kJ/mol. How many kJ energy is needed to convert 1.2 g of Mg atoms in the gaseous state into the Mg^{2+} ion?

45. Compare the elements ${}_{10}Ne$, ${}_{11}Na$ and ${}_{17}Cl$ in terms of their second ionization energies.

46. Which elements have highest second ionization energies in a period? Explain.

47. From which of these isoelectronic elements ${}_{9}F^{-}$, ${}_{11}Na^{+}$ and ${}_{12}Mg^{2+}$ is it more difficult to give an electron?

48. Why is the first ionization energy of an element of group 2A higher than that of an element of group 3A in a period? Explain.

49. Why is the first ionization energy of ${}_{19}K$ lower than that of ${}_{20}Ca$, but the second ionization energy of K higher than that of Ca?

50. What is the electronegativity ? Explain.

51. Does electronegativity have any relationship to ionization energy? Explain.

52. How does electronegativity change in a group and period of the periodic table? Explain.

53. Write down the elements ${}_{12}Mg$, ${}_{14}Si$, ${}_{17}Cl$, and ${}_{9}F$ in order of decreasing electronegativity.

54. By considering the electronegativity values, arrange the following bonds with respect to their ionic character.

I. Si–H II. C–Cl III. C–O IV. Al–C V. Si–C

55. State whether the oxides of the elements ${}_{6}C$, ${}_{14}Si$, ${}_{16}S$, and ${}_{20}Ca$ with oxygen (CO, CO_2, CaO, SiO_2, SO_2 and SO_3) are acidic or basic.

56. What is the most active metal in the 4th period?

57. What is the most active nonmetal in the 3rd period?

MULTIPLE CHOICE QUESTIONS

1. According to which property of the elements did Mendeleev build his periodic table?

A) Their physical states at room conditions
B) Their atomic numbers
C) Their melting and boiling points
D) Their atomic masses
E) Their densities

2. Which of the following statements is/are true for the periodic table?

I. Each noble gas is followed by a metal.
II. Each period ends with a metal.
III. Metalloids are found between the metals and nonmetals.

A) I only B) I and III C) I and II D) II and III E) I, II and III

3. Which of the following statements is/are true for the transition elements?

I. They are all metals.
II. Their electron configuration ends with d – orbitals.
III. They are all in group B.

A) I only B) II only C) I and II D) II and III E) I, II and III

4. Which of the following inferences is/are certainly true for the elements that come before the alkali metals;

I. They are noble gases.
II. They are alkaline earth metals.
III. Their electron configurations end with ns^2np^6.

A) I only B) II only C) III only D) I and III E) II and III

5. If X^{3-} ion has 8 electrons in its 3rd energy level, what is the place of atom X in the periodic table?

A) 1st period, group 4A
B) 2nd period, group 3A
C) 3rd period, group 5A
D) 2nd period, group 3A
E) 3rd period, group 3A

6. Which of the following elements is/are the p–block elements of the periodic table?

I. Nonmetals
II. Metalloids
III. Alkali metals

A) I only B) II only C) I and II D) I and III E) II and III

7. Which of the following statements is/are true for a neutral element X, which has 6 full–filled and 3 half–filled orbitals;

I. It is in the 2nd period
II. It has 5 valence electrons
III. It is in the p – block

A) II only B) I only C) I and III D) II and III E) I, II and III

8. Which one of the following electron configurations is not that of a noble gas?

A) $1s^2$
B) $1s^2\ 2s^2$
C) $1s^2\ 2s^2\ 2p^6$
D) $1s^2\ 2s^2\ 2p^6\ 3s^2\ 3p^6$
E) $1s^2\ 2s^2\ 2p^6\ 3s^2\ 3p^6\ 4s^2\ 3d^{10}\ 4p^6$

9. What is the difference in the number of protons of the halogen in 4th period and the halogen in 3rd period?

A) 1 B) 2 C) 8 D) 18 E) 32

10. Which of the elements $_4X$, $_9Y$, $_{13}Z$ and $_{31}T$ is/are the p–block element(s)?

A) Y, Z and T B) T only C) Y and Z D) X, Z and T E) X only

11. Which of the statements below is/are false for metals?

I. All of them are solid at room temperature.
II. They do not conduct electricity in their molten state.
III. Their oxides are generally basic.

A) I only B) II only C) I and II D) I and III E) II and III

12. Which of the following is a nonmetal?

A) $_{3}Li$ B) $_{4}Be$ C) $_{13}Al$ D) $_{16}S$ E) $_{31}Ga$

13. If a neutral atom gives out an electron, which one of the following never occur?

A) Its atomic radius increases.
B) A cation forms.
C) The number of protons decreases.
D) The number of electrons decreases.
E) The number of shells reduces.

14. When the ion X^- is converted into X^+, which of the below decreases?

I. The ionic radius
II. The number of electrons
III. The number of protons

A) I only B) III only C) I and II D) II and III E) I, II and III

15. The electron configurations of five elements are given below. Which of them has the highest first ionization energy?

A) $1s^2$ B) $1s^22s^22p^5$ C) $1s^22s^22p^6$ D) $1s^22s^22p^63s^23p^64s^1$ E) $1s^22s^22p^63s^23p^6$

16. In a group in the periodic table which of the following term(s) increase(s) from top to bottom?

I. The atomic radius
II. The first ionization energy
III. The number of valence electrons

A) I only B) II only C) I and II D) II and III E) I, II and III

17. In a period of the periodic table which type of element has the greatest first ionization energy?

A) Alkali metals
B) Alkaline earth metals
C) Noble gases
D) Halogens
E) Transition elements

18. Which of the groups given below, has the lowest first ionization energy?

A) Noble gases B) Halogens
C) Earth metals D) Alkali metals
E) Alkaline earth metals

19. Which of the below cause(s) two different elements X and Y, to have different first ionization energies?

I. Difference in their atomic radius
II. Difference in their numbers of protons
III. Difference in their numbers of neutrons

A) I only B) I, II and III C) I and III D) II and III E) I and II

20. Which of the statements below is/are true for S^{2-}, S^0 and S^{6+} of which atomic numbers are 16?

I. S^{2-} has the greatest radius.
II. S^{6+} is the most difficult in giving an electron.
III. S^0 is the easiest in gaining an electron.

A) I only B) II only C) I and II D) I and III E) I, II and III

21. Which one of the following properties does not change from top to bottom in a group of the periodic table?

A) Electronegativity
B) Metallic activity
C) Atomic radius
D) Basicity
E) Number of valence electrons

22. What is the mass percentage of S in the compound formed between $^{27}_{13}Al$ and $^{32}_{16}S$?

A) 27 B) 32 C) 54 D) 64 E) 96

CROSSWORD PUZZLE

ACROSS

2 Scientist who complied a periodic table of 56 elements in order of increasing atomic weight.

3 Amount of energy required to remove an electron from an atom.

5 Special name of group 7A elements.

11 Scientist who developed a table of atomic weights.

13 Special name of group 2A elements.

14 Scientist who determined the atomic number of each elements.

15 Scientist who synthesised transuranic elements (the elements after uranium in the periodic table).

16 Special name of group 8A elements.

DOWN

1 Special name of group 1A elements.

2 Russian chemist who arranged atoms in order of increasing atomic weight.

4 Relative ability of an atom in a molecule to attract the electrons of a covalent bond to itself.

6 Scientist who wrote the first extensive list of 33 elements.

7 Scientist who developed 'triads' (groups of 3 elements with similar properties).

8 Special name of group 3A elements.

9 Scientist who arranged more than 60 elements in order of atomic weights.

12 Scientist who discovered the noble gases.

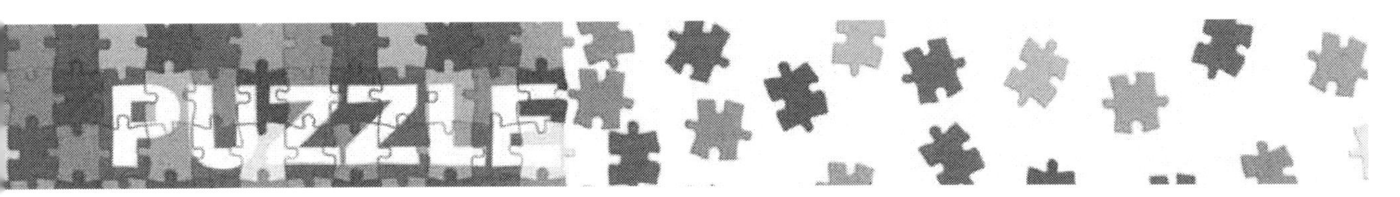

WORD SEARCH PUZZLE

Find the words in the grid. Words can go horizontally, vertically and diagonally in all eight directions.

C	I	Y	S	▪	H	B	K	F	O	C	T	A	V	I	K	K	T	X
S	M	L	K	L	A	N	M	N	X	W	L	Y	€	A	B	H	A	Y
D	Q	P	I	Q	A	B	L	X	T	H	M	T	F	C	L	F	L	L
N	Q	L	X	C	W	T	F	W	T	N	K	D	D	T	F	Q	T	A
A	Q	M	A	N	T	G	I	▪	P	▪	O	R	G	I	N	D	N	T
L	R	K	M	N	M	R	T	M	A	R	V	Y	N	N	€	K	R	I
W	K	Y	N	P	T	O	O	H	N	G	F	I	C	I	C	Q	N	B
I	R	M	X	L	R	H	S	N	R	O	T	K	Y	D	T	K	L	R
N	W	L	M	L	P	U	A	I	I	Y	I	I	T	I	R	N	B	O
N	Y	F	T	I	€	F	€	N	L	G	L	T	R	S	H	M	F	L
K	V	X	R	€	T	V	L	T	I	I	A	L	I	T	B	€	G	U
K	H	I	U	L	D	I	R	C	S	D	Y	T	I	S	N	Q	M	M
G	O	B	N	A	W	I	T	O	B	H	I	B	I	L	N	B	R	T
D	V	K	I	T	T	L	M	B	R	M	R	S	L	V	D	A	N	T
M	X	R	B	T	T	I	K	T	W	O	L	H	R	D	I	T	R	V
I	T	K	T	X	P	D	S	I	A	B	O	R	G	U	R	T	K	T
Y	H	G	K	K	R	N	M	T	H	O	L	O	G	I	N	S	Y	G
I	T	D	Y	G	R	I	N	I	N	O	I	T	A	€	I	N	O	I
R	R	N	U	H	T	M	Q	M	R	U	N	N	€	C	H	P	F	Y

ACTINIDES
AUFBAU
ELECTRONEGATIVITY
GROUP
HALOGENS
IONIZATIONENERGY
LANTHANIDES
MENDELEEV
MEYER
MOSELEY
TRIAD
NEWLANDS
OCTAVE
ORBIT
ORBITAL
PERIOD
SEABORG
TRANSITIONMETALS

NOTES

RADIOACTIVITY

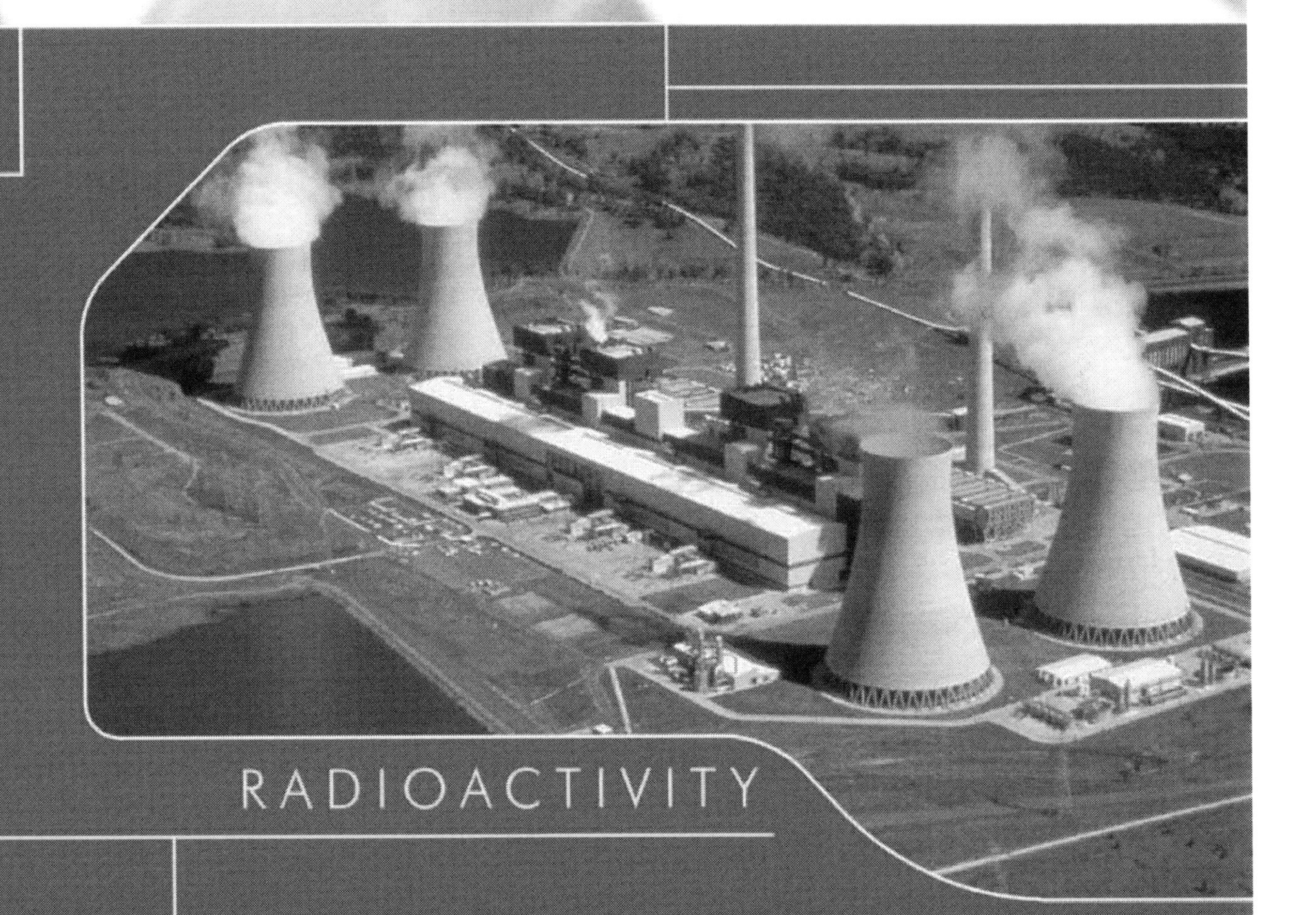

INTRODUCTION

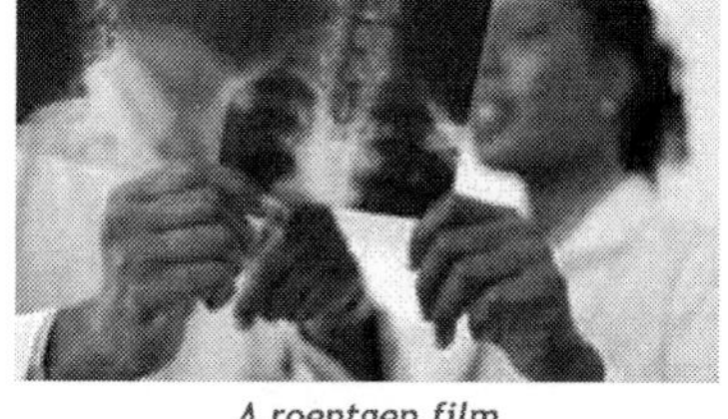

A roentgen film

Almost everything in nature has some small amount of natural radioactivity. We are also bathed in a sea of natural radiation coming from the sun and electrical devices such as cellphones, photocopy machines, x-ray machines. You may not live next to a nuclear power plant, but you are continuously being exposed to radiation.

Traditional chemical reactions occur as a result of the interaction between the valence electrons of atoms. In 1896, Henri Becquerel expanded the field of chemistry to include nuclear changes when he discovered that uranium emitted radiation. Soon after Becquerel's discovery, Marie Skladowska Curie began studying radioactivity and completed much pioneering work on nuclear changes.

Radiation is very harmful in high doses. On the other hand understanding the nature of the different types of radiations, their penetration abilities and ionization power allows us to use radioactive substances safely.

A Comparison of Nuclear and Ordinary Chemical Reactions

Although nuclear and ordinary chemical reactions have a few similarities, they differ in some basic properties.

1. Both types of reactions require activation energy to initiate.
2. Total energy is conserved in both reactions.
3. In chemical reactions, there is no change in the structure of nucleus of an atom. However, in nuclear reactions the structure of nucleus is changed; that is, there is a change in the number of protons and neutrons.
4. The numbers and the types of atoms are conserved in chemical reactions, whereas in nuclear reactions this may not be so.
5. Total mass is conserved in chemical reactions, whereas in nuclear reactions total mass is not conserved.
6. The rate of an ordinary chemical reaction is easily affected by factors such as heat, pressure, concentration and catalyst; the rate of a nuclear reaction, on the other hand, is not affected by any of these factors.
7. The amount of energy released by nuclear reactions is tremendously higher than that of ordinary chemical reactions.

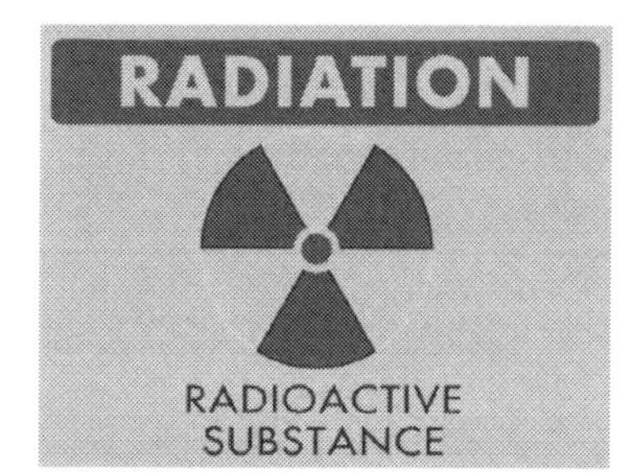

The symbol above is the international symbol representing the risk of radioactivity in an area or a material.

1. RADIOACTIVITY

The spontaneous emission of radioactive rays by an unstable atomic nucleus is called radioactivity. Spontaneously disintegrating atoms are called radioactive atoms and the nuclei of these atoms are unstable.

The stability of the nucleus of an atom depends on the number of protons and neutrons. The nuclei of lighter atoms such as He, C, N and O have the same number of neutrons (N_n) and protons (N_p). Therefore, the ratio between the number of neutrons and the number of protons is 1 ($N_n/N_p = 1$) and such nuclei are stable. There are no atomic nuclei with the same proton and neutron numbers after the element $^{40}_{20}Ca$.

Most radioactive elements have the following properties:

1. The ratio between the numbers of neutrons and protons in a nucleus is greater than 1.5 ($N_n/N_p > 1.5$).
2. The atomic number is bigger than 83.

On the other hand, we can not absolutely say that the isotopes of elements with an atomic number smaller than 83 are all stable. For example, although ^{12}C is $_6$ stable, the isotope $^{14}_{6}C$ is unstable, it is radioactive.

Radioactivity depends on the structure of the nucleus, which contains protons and neutrons.

2. TYPES OF RADIOACTIVE DECAYS

A radioactive substance decays by making mainly three types of emissions. These emissions are named as **alpha, beta** and **gamma**, and represented by symbols α, β **and** ψ respectively. The radiations α, β and ψ emitted by a radioactive substance can have various effects, for instance:

- They can ionize substances.
- Like ordinary light, they can cause substances to phosphoresce and fluoresce.
- They can affect photographic film.
- Substances such as glass, porcelain and ceramics can be colored when they are exposed to radiation. Coloring occurs along the path of radiation.
- Radioactive radiation affects living cells and can cause them to suffer cancer. Also, some inherent illnesses can be caused, and passed from one generation to another.

Now, let us analyze each type of radioactive decay individually.

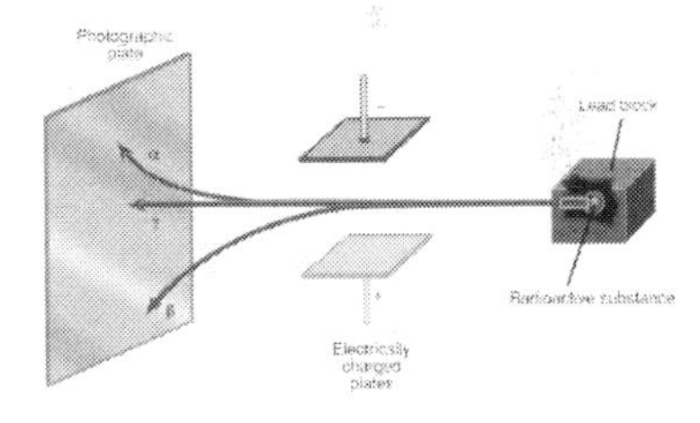

The movement of the radioactive rays in an electric field.

2.1. ALPHA (α) PARTICLES

Schematic representation of alpha particles sign.

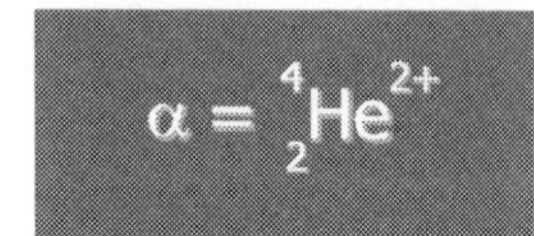

Alpha (α) particles are identical to the nuclei of helium atoms ($^{4}_{2}He$). Their speed is approximately 1/10 of the speed of light. Since they are 2+ charged particles, they are symbolized as $^{4}_{2}He^{2+}$. They are deflected by electric and magnetic fields because of their positive charge.

As they penetrate through matter, alpha particles produce large amount of ions, but their penetrating power is so low that they can easily be stopped by a sheet of paper.

The nucleus of an atom undergoing an alpha decay loses He^{2+} particles: that is, it emits two protons and two neutrons. Therefore, there will be a decrease of two in the atomic number and four in the atomic mass number of that nucleus.

For instance, if an $^{238}_{92}U$ isotope radiates one alpha particle, the nuclear equation can be expressed as shown below.

$$^{238}_{92}U \longrightarrow {}^{234}_{90}Th + \alpha$$

Example 1

The atomic number and atomic mass numbers of each sides are equal to each other in nuclear reactions.

Radioactive $^{230}_{90}Th$ isotope emits one α particle and transmutes to atom X. Find the atomic number, the atomic mass and the symbol of the new element produced.

Solution

The nuclear equation of this radioactive decay can be written

$$^{230}_{90}Th \longrightarrow {}^{A}_{Z}X + \alpha$$

Since the sum of the atomic mass numbers on both sides should be equal, the atomic mass number of X is

$$230 = A + 4 \quad \Rightarrow \quad A = 226$$

And since the sum of the atomic numbers on both sides must also be equal

$$90 = Z + 2 \Rightarrow Z = 88$$

When we look at the periodic table, we see the element Ra with the atomic number 88.

Exercise 1

When the radioactive $^{207}_{82}Pb$ isotope emits two alpha particles, it transmutes into another atom. Find the atomic number and the atomic mass number of that atom.

Answer : 78 and 199.

2.2. BETA (β^-) PARTICLES

Beta (β^-) particles are identical to electrons, moving at approximately the speed of light. Since they are negatively charged particles, they show exactly the same properties as electrons. Therefore, beta particles are represented by either $^{0}_{-1}e$ or β^- symbols. They are lighter and faster than alpha particles, which why their penetrating power is greater than that of alpha particles. For instance, they can pass through aluminum foil two to three millimeters thick. Since beta particles have the opposite charge to alpha particles, they are deflected by electric and magnetic fields in the opposite direction. And since β^- particles have comparatively small masses, these particles are deflected more.

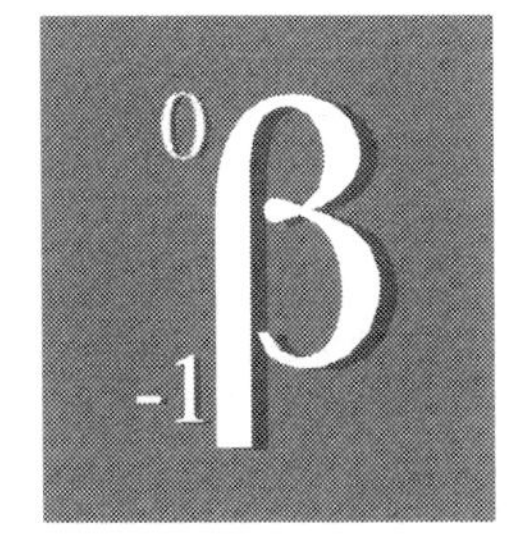

Schematic representation of beta particle sign.

$$\beta^- = {}^{0}_{-1}\beta = {}^{0}_{-1}e$$

An atom undergoing a beta decay, emits an electron ($^{0}_{-1}e$) from its nucleus. This emitted electron is produced as a result of the transmutation in the nucleus of one neutron into one proton.

The equation of this process can be shown thus

$$^{1}_{0}n \longrightarrow {}^{1}_{1}p + \beta^-$$

As a result, for an atom undergoing a β^- decay, the atomic number of the element increases by one, whereas the atomic mass number remains unchanged.

For instance, when the $^{234}_{90}Th$ isotope radiates one beta particle, it transmutes into the $^{234}_{91}Pa$ isotope. The nuclear equation can be expressed as shown below:

$$^{234}_{90}Th \longrightarrow {}^{234}_{91}Pa + \beta^-$$

Example 2

The element $^{218}_{84}Po$ transmutes into the element At by decaying through one β^- radiation. What is the number of protons of At?

Solution

Let us first write the nuclear equation of this decay:

$$^{218}_{84}Po \longrightarrow {}^{A}_{Z}At + \beta^-$$

The sum of the atomic numbers on both sides must be equal. So, the value of Z is

$84 = Z + (-1) \Rightarrow Z = 85$

Since the atomic number is simply the number of protons in an atom, it is equal to 85

Exercise 2

The element of $^{218}_{83}Bi$ transmutes into the element of X by decaying through one β^- radiation. What is the number of protons and the name of element produced?

Answer : 84 and Polonium

2.3 GAMMA (ψ) RAYS

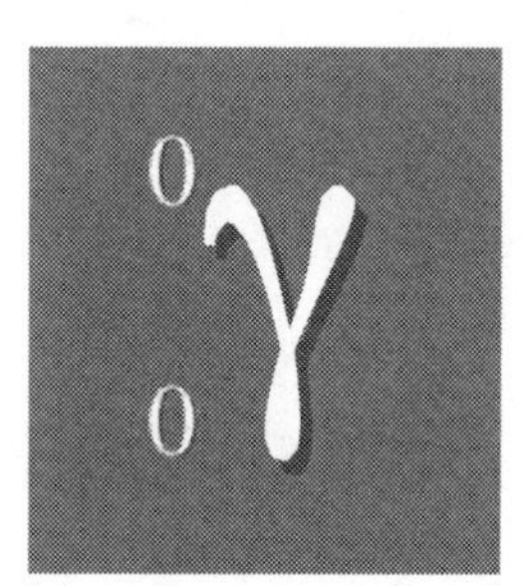

Schematic representation of gamma ray sign.

Gamma (ψ) rays are a type of electromagnetic radiation with a very high energy. Generally, gamma rays are not radiated alone. Some radioactive decays yielding α or β^- particles leave a nucleus in an highly energized (excited) state. Such a nucleus is unstable and denoted by putting an asterix (*) on the upper right corner of the symbol. This unstable nucleus then loses excess energy in the form of gamma rays to become a more stable nucleus with lower energy.

The penetrating power of gamma rays is approximately one hundred times greater than that of beta rays. For example, gamma rays can pass through a lead block several centimeters thick. They can only be stopped by concrete blocks of two to three meters thick.

Since gamma rays are electrically neutral particles, they are not deflected by electric or magnetic fields and they do not have an ionizing effect.

The mass and charge of gamma rays are accepted as zero. Therefore, the atomic mass number and the atomic number of an atom remain unchanged when it radiates a gamma (ψ) ray.

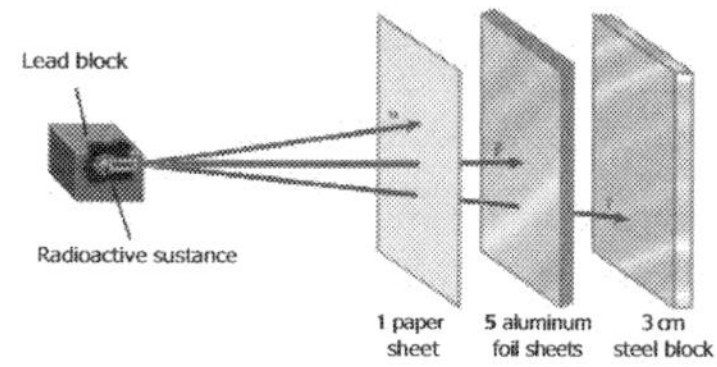

Comparision of penetrating powers of the α, β, and ψ rays.

For example, when the element $^{234}_{92}U$ transmutes into $^{230}_{90}Th$ by radiating 1α ray, gamma (ψ) rays are also produced.

The nuclear equation of this process is shown below;

$$^{234}_{92}U \longrightarrow {}^{230}_{90}Th^* + \alpha$$

$$^{230}_{90}Th^* \longrightarrow {}^{230}_{90}Th + {}^{0}_{0}\psi$$

2.4 POSITRON EMISSION

In some radioactive decays, one proton is transformed into one neutron and a positively charged particle with the same mass as a beta (β^-) particle (or electron) is produced. Emission of this positively charged particle is known as **positron** emission. Positrons (antielectrons) are symbolized as β^+ or $^{0}_{+1}e$. The nuclear equation of the formation of a positron particle is:

$$^{1}_{1}p \longrightarrow {}^{1}_{0}n + {}^{0}_{+1}e$$

As a result of a positron emission the atomic number of an element decreases by one, whereas the atomic mass remains unchanged

The following nuclear equations can also be given as examples of positron emissions:

$${}^{23}_{12}Mg \longrightarrow {}^{23}_{11}Na + {}^{0}_{+1}e$$

$${}^{30}_{15}P \longrightarrow {}^{30}_{14}Si + {}^{0}_{+1}e$$

Example 3

A radioactive isotope ${}^{38}_{19}K$ transmutes into Ar by decaying through one β^+ emission. Find the atomic mass number of Ar.

Solution

The nuclear equation can be written thus:

$${}^{38}_{19}K \longrightarrow {}^{38}_{18}Ar + {}^{0}_{+1}\beta$$

The atomic mass number of Ar remains at 38.

Exercise 3

A radioactive element decays through one positron (β^+) emission and gives potassium, ${}^{39}K$. Find the element decayed.

Answer : ${}^{39}_{20}Ca$

2.5 ELECTRON CAPTURE

Electron capture has the same effect as positron emission. In this process, an electron from an inner energy shell (especially the 1s orbital) is captured by the nucleus. The captured electron is used to convert a proton into a neutron. This conversion can be shown thus:

$${}^{1}_{1}p + {}^{0}_{-1}e \longrightarrow {}^{1}_{0}n$$

The atomic number of the nucleus decreases by one, but its atomic mass number remains unchanged. The energy level left empty by the captured electron is filled by an electron from a higher energy level. When the electron drops from the higher energy level to the lower one, the energy is emitted as the difference between these two levels as in the form of electromagnetic radiation (X-radiation).

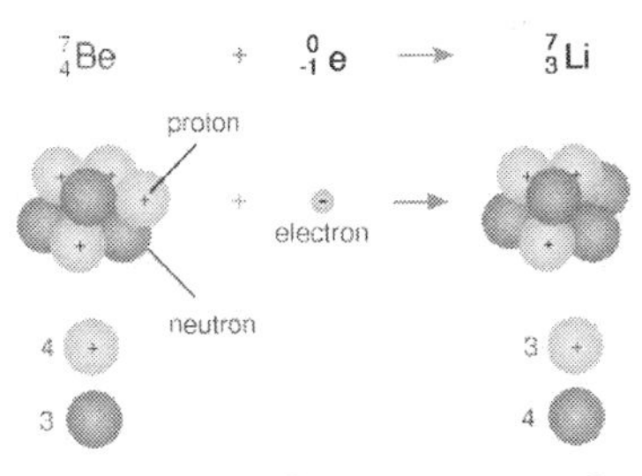

Transmutation of ${}^{7}_{4}Be$ isotope into the ${}^{7}_{3}Li$ isotope through an electron capturing.

The following examples can be given for electron capture:

$${}^{106}_{47}Ag + {}^{0}_{-1}e \longrightarrow {}^{106}_{46}Pd$$

$${}^{202}_{81}Tl + {}^{0}_{-1}e \longrightarrow {}^{202}_{80}Hg$$

Example 4

$$^{56}_{26}Fe + ^{0}_{-1}e \longrightarrow X$$

By considering the nuclear equation given above, find the number of neutrons in X.

Solution

The sum of the atomic numbers on both sides must be equal in the nuclear equation.

$$^{56}_{26}Fe + ^{0}_{-1}e \longrightarrow ^{A}_{Z}X$$

Therefore, the atomic number of X will be

$26 + (-1) = Z \Rightarrow Z = 25$

Similarly, the sum of the atomic mass numbers on both sides must also be equal. Therefore, the atomic mass number of X is

$56 + 0 = A \Rightarrow A = 56$

Finally, the number of neutrons of the element X produced is equal to A – Z, so

$56 - 25 = 31$

Exercise 4

An isotope of beryllium, $^{7}_{4}Be$, captures an electron. Which element would be produced?

Answer : Lithium, Li

2.6 NEUTRON EMISSION

Neutron emission occurs by the removal of a neutron from a nucleus. However, it is very difficult to trace a neutron emission, since it occurs very rarely and at a high velocity. A new element is not formed as a result of this decay, but an isotope of the original element is formed. That is, after a neutron emission the atomic number of the element remains unchanged whereas the atomic mass number decreases by one. The following nuclear equation is an example for a neutron emission:

$$^{87}_{36}Kr \longrightarrow ^{86}_{36}Kr + ^{1}_{0}n$$

$$^{5}_{2}He \longrightarrow ^{4}_{2}He + ^{1}_{0}n$$

Transmutation of a radioactive nucleus, $^{5}_{2}He$, into its isotope by neutron emission.

Example 5

A radioactive nucleus ^5_2He transmutes into one of its isotopes through a neutron emission. What is the neutron/proton ratio in the new nucleus produced?

Solution

First of all, let us write the nuclear equation of this emission:

$$^{5}_{2}He \longrightarrow {}^{A}_{Z}He + {}^{1}_{0}n$$

The atomic mass number of this isotope is

$5 = A + 1 \Rightarrow A = 4$

Therefore, the number of neutrons present in this nucleus is $A - Z = 4 - 2 = 2$.

Thus, the n/p ratio in this nucleus is $2/2 = 1$.

The table given below summarizes the changes in the atomic numbers and the atomic mass numbers of an element as a result of the nuclear radiations described.

Types of decay	Symbol	Chemical symbol	Changes		Example
			Atomic number	Atomic mass number	
Alpha	α	$^{4}_{2}He$	Decreases by 2	Decreases by 4	$^{238}_{92}U \longrightarrow {}^{234}_{90}Th + \alpha$
Beta	β^-	$^{0}_{-1}e$	Increases by 1	No change	$^{234}_{90}Th \longrightarrow {}^{234}_{91}Pa + {}^{0}_{-1}e$
Gamma	ψ	$^{0}_{0}\psi$	No change	No change	$^{230}_{90}Th^* \longrightarrow {}^{230}_{90}Th + {}^{0}_{0}\psi$
Positron	β^+	$^{0}_{+1}e$	Decreases by 1	No change	$^{23}_{12}Mg \longrightarrow {}^{23}_{11}Na + {}^{0}_{+1}e$
Neutron	n	$^{1}_{0}n$	No change	Decreases by 1	$^{87}_{36}Kr \longrightarrow {}^{86}_{36}Kr + {}^{1}_{0}n$

Some important radioactive emissions and changes in the nucleus of an atom.

Antoine Henri Becquerel
(1852–1908)

Becquerel was a French physicist who discovered radioactivity. For this discovery and other valuable studies, he shared the Nobel Physics Prize with Pierre and Marie Curie in 1903.

Marie Curie
(1867–1934)

Marie Skladowska shared the Nobel Prize in Physics in 1903 with Henri Becquerel. Marie studied radium, and finally succeeded in isolating pure radium in 1910. For this achievement she won the 1911 Nobel Prize in Chemistry. After studying radioactivity for such a long period of time without any self-protection, she developed cancer and died in 1934.

3. NATURAL RADIOACTIVITY

Henry Becquerel's pioneering discovery of radioactivity on 1 March 1896 led to the production of several useful technological instruments for the benefit of mankind.

In 1895 X-rays were discovered by Wilhelm Roentgen. At first Henry Becquerel (1852 – 1908) thought that X-rays were related to his studies and he imagined that fluorescent materials would produce X-rays. He then designed experiments to prove this.

He observed that there were no X-rays emitted by fluorescent materials.

Later, Becquerel used an uranium salt, uranyl sulfate, $K_2UO_2(SO_4)_2 \cdot 2H_2O$, in his experiment.

Becquerel had discovered an important property of uranium salt. He had found that uranium salt emitted radiation which penetrated black paper irregardless of sunlight.

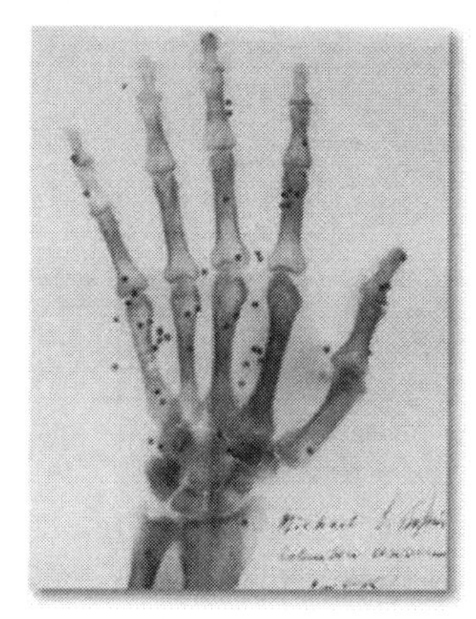

Just three months after their discovery in Germany, X-rays were already being used here to examine the hand of a New York attorney who had been accidently shot with a shotgun.

That was the story of how Becquerel discovered radioactivity. The rest of this story is about the Curie family.

Becquerel continued his studies, but he had limited himself by using uranium as the only source of rays. That is why the Curies (Pierre and Marie) took a big step forward, but Becquerel could not.

At the end of long and hard days, they isolated a new element. From **"pitchblende"**, an uranium ore, they obtained a new element which radiates rays similar to uranium. They named this new element **"polonium"** to honor the memory of Poland, Marie Curie's homeland. This discovery led to the discovery of **"radium"** which made the Curies famous. With the discoveries of these new radioactive elements, the number of such elements reached four. They were **uranium, thorium, polonium** and **radium.**

Because of ionizing rays emitted by these elements, Marie Curie named such elements, "radioactive" elements. This is the origin of the term, radioactivity.

Important Dates in the History of Radioactivity

1895 X-rays were discovered. (W.C. Roentgen)

1896 The radioactivity of uranium was discovered. (H. Becquerel)

1897 Electrons were discovered. (J.J. Thomson)

1898 The elements polonium and radium were discovered. (P. and M. Curie)

1899 α rays emitted by thorium were discovered. (E. Rutherford)

1900 Alpha and Beta rays emitted by radium were discovered. (P. Curie)

1900 Gamma rays were discovered. (P.U. Villard)

1903 Hilgartner was used external radiotherapy for the treatment of retinoblastoma(eye cancer) and first case was published in the Texas Medical Journal.

1905 Mass-Energy equality was defined as $E = mc^2$ (A. Einstein)

1910 The idea of radioactive isotopes was proved. (F. Soddy)

1919 The first artificial nucleus transmutation was achieved. (E. Rutherford)

1932 Neutrons were discovered. (J. Chadwick)

1934 β^+ emission and the first artificial radioactivity were achieved. (J.F and I. Joliot – Cruie)

1938 Nuclear division was acheived. (O. Hahn and F. Strassmann)

1942 The first nuclear fission reaction was achieved. (E. Fermi)

1945 The first atomic bombs were exploded in Hiroshima and Nagasaki.

DAILY DISPATCH

ATOM BOMB HITS JAPS

St. Cloud Man Named New Head of Dental Group

1947 The Atomic Energy Commission begins work on a report investigating peaceful uses of nuclear energy.

1951 First electricity was generated from atomic power at EBR-1 Idaho National Engineering Lab, Idaho Falls.

1954 The world's first nuclear power plant that generated electricity for commercial use was officially connected to the Soviet power grid at Obninsk, Kaluga Oblast, Russia.

1955 The first U.S. town Arco, Idaho, with a population of 1,000, is powered by nuclear energy.

1986 The worst nuclear accident was occurred in Chernobyl nuclear reactor near Kiev Soviet Union and caused nearly 30,000 cancer-related deaths over a 50 year period.

THE PIONEERS

Wilhelm Conrad Roentgen
(1845–1923)

Roentgen, a German scientist, first discovered X-rays. When he was studying on 8 November 1895, he noticed that the barium cyanide crystals on the table were emitting light. He thought that the reason for this was that the invisible rays coming from the crockes tube he was using excited the barium cyanide crystal giving fluorescent abilities. Roentgen could not explain the structure of these rays and gave them the name of X-rays, meaning unknown rays. Thus, he won the first Nobel Prize of Physics in December 1901.

4. ARTIFICIAL RADIOACTIVITY

Some stable nuclei, which are not radioactive, can undergo radioactive changes by bombarding them with various particles. These bombarded nuclei are transmuted into new nuclei as a result of the bombardment. This process is called **artificial radioactivity.**

The non-radioactive element, $^{17}_{8}O$ was first produced artificially by Ernest Rutherford in 1919. In order to achieve this, he bombarded stable $^{14}_{7}N$ nucleus with α–particles emitted by radium and polonium.

Following this theory, the transformation of one element into another one was realized. The nuclear equation of this artificial radioactive reaction is illustrated below;

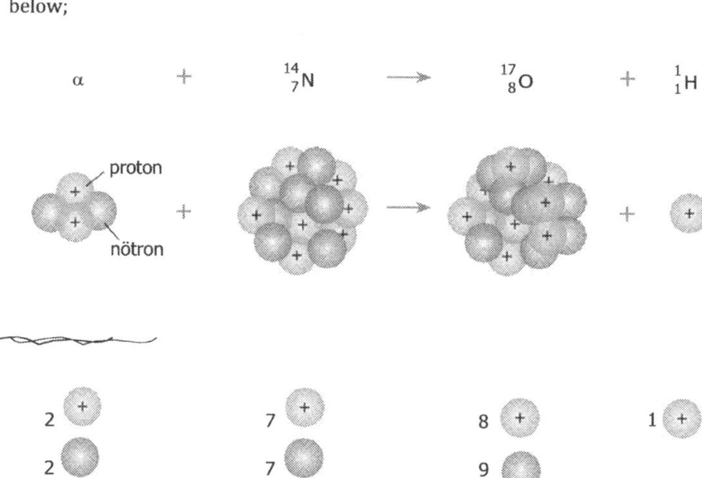

Bombardment of a non-radioactive isotope $^{14}_{7}N$ by an α particle.

The discovery of neutron particles was based on artificial nuclear reactions. James Chadwick bombarded the nuclei of $^{9}_{4}Be$ atoms with α–particles and obtained neutron particles in 1932.

$$^{9}_{4}Be + \alpha \longrightarrow {}^{12}_{6}C + {}^{1}_{0}n$$

THE PIONEERS

Ernest Rutherford
(1871–1937)

Rutherford discovered the nucleus of the atom and its properties. He was also the first scientist to succeed in the transmutation of an element into the isotope of another element. For his study of radioactive elements he won the 1908 Nobel Prize in Chemistry.

Example 6

When a non-radioactive $^{235}_{92}U$ isotope is bombarded with α particles to produce the isotope $^{238}_{93}Np$, an X particle is emitted. What is the emitted X particle?

Solution

The equation of this nuclear reaction is written thus:

$$^{235}_{92}U + \alpha \longrightarrow ^{238}_{93}Np + ^{A}_{Z}X$$

The sum of the atomic numbers on both sides should be equal:

$91+2 = 92+Z \Rightarrow Z = 1$

Similarly, the sum of the atomic mass numbers on both sides should be equal:

$235 + 4 = 238 + A \Rightarrow A = 1$

Thus, the particle X, having the atomic number and the atomic mass number of 1, is simply a proton ($^{1}_{1}p$).

Exercise 5

When an atom of $^{106}_{46}Pd$ is bombarded with α particles, its nucleus captures one α particle and emits one proton and transmuted into Ag. What is the number of neutrons of the Ag atom produced?

Answer : 62

4.1. NUCLEAR FISSION

When heavier nuclei are bombarded by slow neutrons, the nuclei of lighter elements are formed. Besides the energy released, several neutrons are emitted. The disintegration of a heavier nucleus into lighter nuclei by neutron bombardment is called **nuclear fission (nuclear division).**

In 1938, Otto Hahn and Fritz Strassman of Germany proved that the bombardment of uranium nuclei with neutrons produces several lighter and stable nuclei, each having about equal sizes. The produced nuclei are the radioisotopes of the lighter elements such as barium (Ba) and cerium (Ce). Thus, the foundations of a new useful method for the production of huge amount of energy were established.

The element uranium is the element used for almost all fission processes. It has two natural isotopes. One of them is ^{238}U which, constitutes 99.3% of uranium ore, and the other is ^{235}U, which constitutes 0.7% of uranium ore. Fissionable nuclei such as ^{235}U and ^{239}Pu are called **fissile**. Nuclear fission reactions occur

Nuclear Reactor

A nuclear reactor is a device in which nuclear chain reactions are initiated, controlled, and sustained at a steady rate. Nuclear reactors are used for many purposes, but the most significant current uses are for the generation of electrical power and for the production of plutonium for use in nuclear weapons. Currently, all commercial nuclear reactors are based on nuclear fission. The amount of energy released by one kg ^{235}U is equal to the energy from the combustion of 3000 tons of coal or the energy from an explosion of 20,000 tons of TNT (Trinitrotoluene, called commonly dynamite).

This nuclear power station has four nuclear reactors, which are inside the black square-shaped buildings.

through the slow neutron bombardment of fissile nuclei.

$$^{235}_{92}U + ^{1}_{0}n \longrightarrow ^{92}_{36}Kr + ^{141}_{56}Ba + 3^{1}_{0}n$$

In addition to the huge amount of energy released by nuclear fission reactions, another important result of such reactions is that more neutrons are produced than the number of neutrons used to bombard. The produced neutrons may also strike other ^{235}U isotopes and causes new fissions. The new nuclear fission reactions also produce neutrons with huge amounts of energy, and so on. This continuous process is said to be the **atomic bomb,** and is the basic principle of nuclear reactors.

Enrico Fermi
(1901–1954)

Fermi was an Italian physicist who emigrated to the USA in 1939 and continued his studies there. Fermi produced a nuclear chain fission reaction by bombarding the isotope ^{238}U with neutrons. For his this achievement he won the 1938 Nobel Prize in Physics.

This chain reaction was firstly achieved by Enrico Fermi on September 2, 1942 in Chicago.

The first test of the atomic bomb was carried out on July 16, 1945 in the New Mexico desert.

After this successful test, the first atomic bomb was dropped on Hiroshima during World War II, on August 6, 1945. The second atomic bomb was dropped three days later on Nagasaki, leading to the end of the World War II.

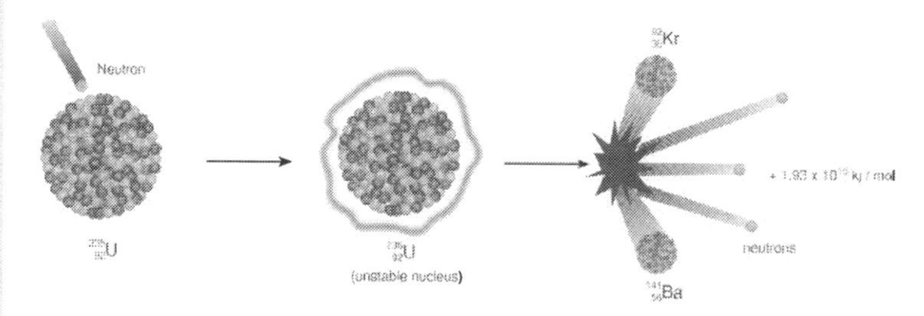

Disintegration of an $^{235}_{92}U$ nucleus into ^{92}Kr and ^{141}Ba as a result of neutron bombardment.

READING — The Atomic Bomb

The main feature of the atomic bomb is that the fission reaction is carried out in a very short time. In the atomic bomb two types of substances are used, a natural one, uranium (^{235}U), and an artificial one, plutonium (^{239}Pu). One of the basic problems in the production of an atomic bomb is obtaining the necessary ingredients. The uranium isotope ^{235}U is found in trace amounts of natural sources of ^{238}U. The isotope ^{235}U used in an atomic bomb has to be as pure as possible. Therefore, the isotope ^{235}U must be separated from the isotope ^{238}U. The isotope ^{239}Pu does not occur naturally. It is produced from the isotope ^{238}U in nuclear power stations.

An atomic bomb therefore has a core containing uranium or plutonium at a center. For a nuclear explosion the amount of the core has to be greater than the critical mass that may explode itself. Therefore, the explosive core is divided into different portions and placed into the bomb. When the atomic bomb is going to be ignited, these portions have to come together and form a spherical shape. In order to form the spherical shape, trinitrotoluene (TNT) is used. First, TNT is exploded and then the nuclear explosives come together and the main explosion takes place.

The first studies of the atomic bomb began under the leadership of R. J. Oppenheimer in late 1942. The team of scientists working at Los Alamos, in New Mexico state, produced the first result three years later.

The atomic bomb was first tested in a desert area near Alamogordo, New Mexico, on July 16, 1945. The power of the explosion was approximately equal to the effect of 20,000 tones of TNT. After achieving successful results, it was decided to use the atomic bomb on two important cities in Japan.

The first atomic bomb was dropped on Hiroshima the morning on August 6, 1945. by a bomber plane named "Enola Gay". The first effect of the explosion occurred in one tenth of a second. It was a very bright light which caused blindness and 300,000°C heat. It burnt everything within about a three kilometer diameter area. An 1,800 km/h shock wave from the explosion destroyed everything. The major and permanent effect was the radioactive fallout which started several minutes later. This terrifying bomb, which ruined Hiroshima in a few seconds, caused approximately 80,000 deaths and 100,000 wounded.

The second atomic bomb was dropped on Nagasaki on August 9, 1945. In this occasion, the number of dead was not so high as in Hiroshima because the people in Nagasaki had been warned previously. However, diseases and deaths from nuclear radiation have continued ever since August 15, 1945.

Volunteers trying to rescue their relatives in the ruined area had unknowingly exposed themselves to nuclear radiation.

The number of deaths caused by the nuclear radiation was much more greater than the number of deaths caused by the very high heat, shock and the fallout at the time of explosion.

The atomic bomb named 'Little boy" which was dropped on Hiroshima weighed 4,500 kg, had a diameter of 75 cm and a length of 3.5 m.

The atomic bomb named "Fat man" which was dropped on Nagasaki weighed 5,000 kg, and had a length of 4 m.

Hiroshima was one of the biggest cities in Japan. This picture shows the remains of the city a few days after the atomic bomb explosion.

4.2. NUCLEAR FUSION

Nuclear fusion reactions occur naturally in the sun for about five billion years. These reactions are the only source of energy sustaining life on our planet.

The combination of two or more lighter nuclei to form a heavier nucleus is called **nuclear fusion**. The amount of energy released in fusion reactions is greater than the amount of energy released in fission reactions. However, a huge amount of activation energy (such as an atomic bomb explosion) is needed to initiate nuclear fusion reactions.

The simplest nuclear fusion reaction is the combination of the isotopes of hydrogen (deuterium and tritium) to form the heavier nucleus of helium.

$$^{2}_{1}H + ^{3}_{1}H \longrightarrow ^{4}_{2}He + ^{1}_{0}n + \text{energy} \qquad ^{2}_{1}H + ^{2}_{1}H \longrightarrow ^{4}_{2}He + \text{energy}$$

The nuclei of deuterium and tritium have to get very close to each other to initiate nuclear fusion reactions. Since atomic nuclei repel each other, the nuclei must have very high energy to overcome the repellent forces between them. This process is possible at temperatures above 100 million degrees centigrade. At this temperature, atoms are separated into their components (nuclei and electrons) and a plasma state is attained. Hence, substances are in the form of positively charged nuclei floating electron clouds. The energy sources of the sun and stars are samples of fusion reactions. The most important advantage of fusion reactions with respect to fission reactions is the abundance of hydrogen and its isotopes in nature. Nuclear fusion is the basis of the hydrogen bomb. An atomic bomb was first exploded to initiate the reactions in the hydrogen bomb. The effects of the atomic bomb, high temperature and pressure, started the nuclear fusion reaction. In other words, the igniter of hydrogen bomb is an atomic bomb. The power of a hydrogen bomb is about 1000 times greater than that of an atomic bomb.

The 'Tokamak' device was invented in 1960, and several nuclear fusion experiments have been carried out in it. Tokamaks are containers which have a very high quality vacuum and within which there are the strongest known magnetic fields.

5. THE RATE OF RADIOACTIVE DECAY

The changing of radioactive elements into other elements through radioactive emissions is called either **radioactive decay** or **radioactive disintegration**. By using radioactive rays, it is possible to detect whether a substance is radioactive or not. There are several methods to detect the types of radiations and their intensities. The most commonly used device to check the intensity of radioactivity is the **Geiger–Müller counter**.

The radioactivity of each element is not the same. Each radioactive element undergoes a characteristic rate of decay. It is the intensity of this decay which is determined by a Geiger–Müller counter.

The rate of radioactive decay of an element is the number of atoms emitting a radioactive ray per a unit time. The rate of decay is directly proportional to the initial amount of substance and the structure of the nuclei. On the other hand, the rate of decay is independent of the physical and chemical properties of a radioactive atom. Temperature does not affect the rate of decay. The rate of radioactive decay of a nucleus is a criterion for the stability of the nucleus. In order to compare the stabilities of different radioactive substances the rate of radioactive decay is generally expressed as the **half-life**.

The **half-life** of a radioactive decay is the period of time required for half of the initial amount of the substance to disintegrate. The shorter the half-life of a radioactive decay, the higher the rate of radioactive decay and the more radioactivity. The half-life is the characteristic property of each element.

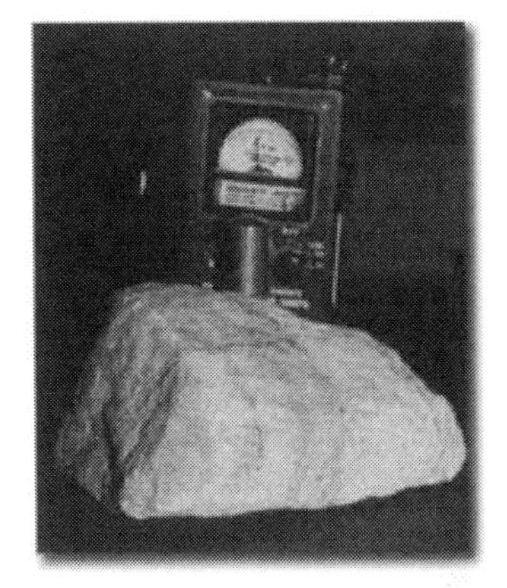

In order to check the radioactivity of a rock, a Geiger-Müller counter is used.

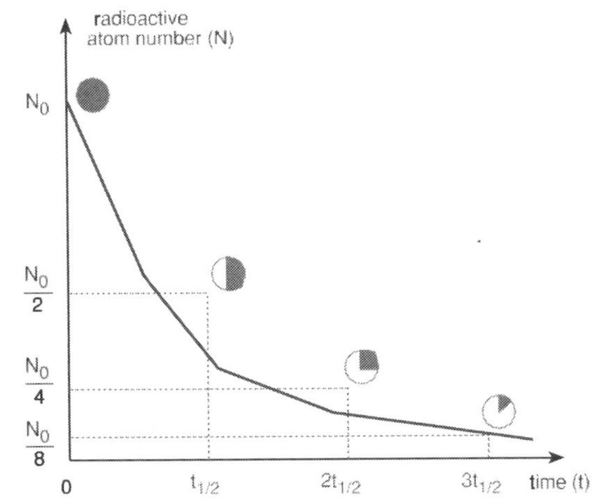

The changing number of atoms of a radioactive isotope with time.

Element	Half-life
$^{214}_{84}Po$	$1.64 \cdot 10^{-4}$ seconds
$^{13}_{8}O$	$8.7 \cdot 10^{-3}$ seconds
$^{80}_{35}Br$	17.6 minutes
$^{28}_{12}Mg$	21 hours
$^{222}_{86}Rn$	3.823 days
$^{32}_{15}P$	14.3 days
$^{234}_{90}Th$	24.1 days
$^{35}_{16}S$	88 days
$^{3}_{1}H$	12.26 years
$^{90}_{38}Sr$	28.1 years
$^{226}_{88}Ra$	$1.60 \cdot 10^{3}$ years
$^{14}_{6}C$	5730 years
$^{40}_{19}K$	$1.25 \cdot 10^{9}$ years
$^{238}_{92}U$	$4.51 \cdot 10^{9}$ years

The half-lives of some radioactive elements.

This graph also shows two important points about half-life.

1. The half-life does not depend on the initial amount of the substance.
2. The half-life is not the period of time required to disintegrate all of the initial quantity of the substance completely.

In the calculations of half-lives, the expressions

$$m = \frac{m_0}{2^n} \qquad \text{where} \quad n = \frac{t}{t_{1/2}}$$

are used. In these expressions;

m_0 = initial mass

m = amount of substances remained. n = the number of half-lives

t = time passed $t_{1/2}$ = half-life.

Now, let us solve some examples related to radioactive decay.

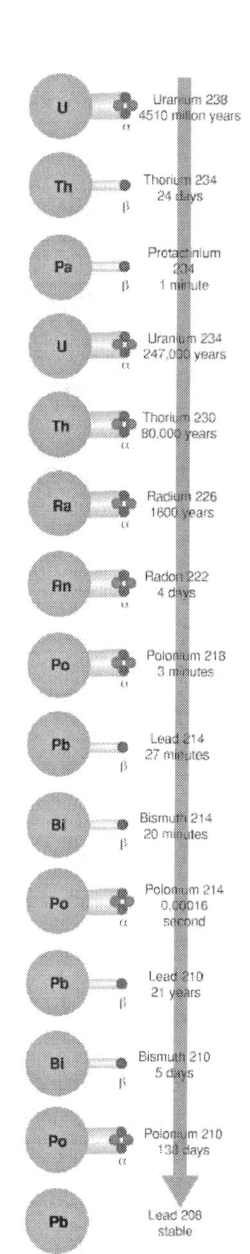

The half-lives of the isotopes found in the uranium decay series.

Example 7

How many grams of two moles ^{212}Pb will remain undecayed after 33 hours if its half-life is 11 hours ?

Solution

There are two ways to solve this problem.

1st Way

2 moles of ^{212}Pb are equal to 424 g, so m_0 = 424 g.

the number of half-lives in 33 hours in the formula of

$$n = \frac{t}{t_{1/2}} \Rightarrow n = \frac{33}{11} \Rightarrow n = 3$$

$m = \frac{0}{2^n}$ ~~is~~ used to find the amount of mass remaining:

$$m = \frac{m_0}{2^n} \Rightarrow m = \frac{424}{2^3} \Rightarrow m = \frac{424}{8} \Rightarrow m = 53\ g$$

2nd Way

The initial mass of ^{212}Pb (424 g) will be halved 3 times. Since after each halving, the amount of ^{212}Pb left will be half of the previous amount, after the first half-life 212 g, after the second half-life 106 g, after the third half-life 53 g of the substance will remain unchanged.

Initial mass		1st		2nd		3rd
424	⇒	212	⇒	106	⇒	53

53 g of ^{212}Pb will remain undecayed at the end of 33 hours.

Example 8

The amount of an 80 grams sample of $^{210}_{83}Bi$ remaining after 20 days is 5 grams. What is the half-life of $^{210}_{83}Bi$ in days?

Solution

Since the initial quantity of the substance is 80 g, after the first half-life 40 g, after the second half-life 20 g, after the third half-life 10 g and after the 4th half-life 5 g of the substance will remain unchanged.

The initial mass of $^{210}_{83}Bi$ will be halved 4 times.

If 20 days pass for 4 decay period
x days pass for 1 decay period

x = 5 days.

The half-life of $^{210}_{83}Bi$ is 5 days.

Exercise 6

30 grams of a radioactive substance decayed after 4 half-lives. What was the initial mass of this radioactive substance?

Answer : 32 g

Exercise 7

87.5% of a radioactive substance decays in one year. What is its half-life in months?

Answer : 4 months

6. THE EFFECTS OF RADIATION ON LIVING ORGANISMS

Radiation is known to be a danger, but this is not totally correct. The existence of radiation has been known since ancient times. On the other hand, through technological and industrial development, the purification of uranium and its widespread usage caused an increase in the effects of radiation.

Many sources of radiation already exist in our world. In fact, electrical devices such as the TV, cellphones, photocopy machines, x-ray machines are common examples. Nuclear reactions can be shown to be one of the important sources of radiation. In nuclear tests (atomic and hydrogen bombs) some radioactive substances are produced and although the reactions stop, the radiation continues.

Another source of radiation is space. As we know the energy of our sun and all other stars form nuclear fusion reactions. Not only heat and light but also nuclear radiation come to the Earth from space. This nuclear radiation is called "cosmic rays" or "cosmic radiations". The earth's ozone layer usually absorbs these types of radiations. However, a very small quantity of cosmic rays reaches the Earth`s surface. Briefly, it is not possible to get rid of radiation.

In space, radioactive rays have the enormous speed of about 300,000 km per second. Due to this, they can easily enter our body and damage our cells. Moreover, these rays may also change the chemical structure of our cells. In fact, electrically charged rays can ionize and decompose some molecules in cells in only one thousandth of a second. In addition, they can affect other cells found in the surroundings of these cells and may disrupt their physiological functions. When these types of cells are exposed to radiation they either die or lose their function. In reality, the death of only a small amount of cells is not so important because in normal conditions the death of some cells results in the rebirth of new ones. However, the sudden death of a large number of cells is very dangerous.

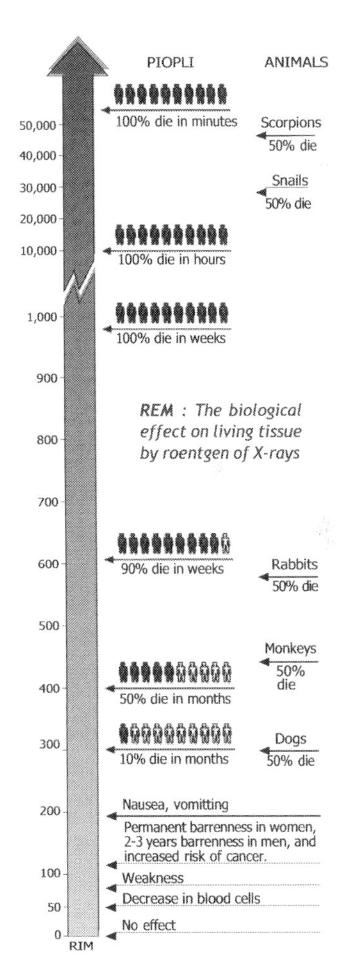

The relationship between the sensitivities of human beings and animals.

The most durable creatures in the face of radioactivity are scorpions.

It has been found that the effects of radiation on some vital tissues (bone marrow, spleen, blood and reproduction cells) are seen more quickly than on other tissues. Because these tissues grow faster than others, any type of damage in the cells can be easily and quickly transferred into new cells. All cells can be affected by this chain reaction. This can result in tumors, which explains the carcinogenic effect of radiation.

READING — Why Do We Need Protection From The Sun?

The sun emits a series of light rays composed of different elements: cosmic rays, gamma rays, X-rays, ultraviolet (UV) (consisting of UVA, UVB and UVC), visible light, infrared (IR) and radio waves. Filtered by the atmosphere, two-thirds of these rays reach Earth. Cosmic radiation, gamma rays, X-rays and UVC rays incompatible with life do not reach the surface. Of the rays which do reach us, only UVA, UVB, visible light and infrared have an effect on our bodies. These rays have beneficial effects: UVB rays promote the synthesis of vitamin D, essential for calcium fixation in the bones, visible light has an anti-depressant effect and infrared has a heating effect which raises skin temperature (an alarm signal to avoid sunburn). On the other hand, in the event of over-exposure, UVA and UVB rays can be particularly harmful. In the short term, they may cause sunburn and trigger photosensitive reactions (pathological skin symptoms linked to the interaction in the skin of an external agent and the sun). Over longer periods, UVA and UVB are responsible for skin ageing and, above all, the occurrence of skin cancers. Conclusion: over-exposure to the sun is dangerous!

The most important danger is the changes in the DNA structure of a cell. Any change in the structure of DNA causes a change in the structure of chromosomes, which can cause the loss of information stored there. By the formation of a new genetic structure, a new genotype (not similar to the parents) of offspring appears. This change is called **'mutation'**.

After mutation in reproductive cells, the changes caused by radiation are transferred to new generations.

Some radioactive isotopes used in medicine for treatment purposes	
Isotope	**Uses**
${}^{11}_{6}C$	Brain tomography
${}^{32}_{15}P$	Treatment of eye tumors
${}^{59}_{26}Fe$	Anaemia
${}^{24}_{11}Na$	Blood circulation disorders
${}^{67}_{31}Ga$	Treatment of lung tumors
${}^{75}_{34}Se$	Pancreatic disorders
${}^{98}_{43}Tc$	Treatment of bone marrow, kidney, liver and brain

Some medical radioactive isotopes and their usage.

READING Carcinogens

Hundreds of chemicals are capable of inducing cancer in humans or animals after prolonged or excessive exposure. There are many well-known examples of chemicals that can cause cancer in humans. The fumes of the metals cadmium, nickel, and chromium are known to cause lung cancer. Vinyl chloride causes liver sarcomas. Exposure to arsenic increases the risk of skin and lung cancer. Leukemia can result from chemically induced changes in bone marrow to exposure to benzene and cyclophosphamide, among other toxicants. Other chemicals, including benzo[a]pyrene and ethylene dibromide, are considered by authoritative scientific organizations to be probably carcinogenic in humans because they are potent carcinogens in animals. Chemically-induced cancer generally develops many years after exposure to a toxic agent. A latency period of as much as thirty years has been observed between exposure to asbestos, for example, and the incidence of lung cancer.

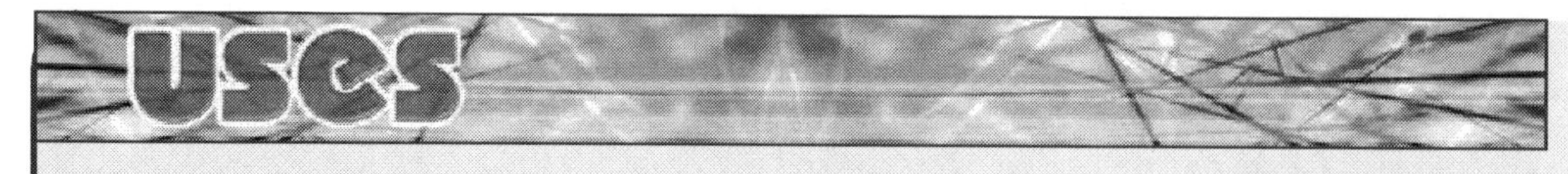

Some Uses of Radioactivity

Smoke Detectors

Smoke alarms contain a weak source made of Americium-241.

Alpha particles are emitted from the alarm, which ionize the air so that it conducts electricity and a small current flows.

If smoke enters the alarm, this absorbs the α particles, the current reduces, and the alarm sounds.

Thickness Control

In paper mills, the thickness of the paper can be controlled by measuring how much beta radiation passes through the paper to a Geiger counter.

The counter controls the pressure of the rollers to give the correct thickness.

With paper, plastic or aluminium foil, β^- rays are used, because alpha will not go through paper.

An α source is chosen that has a long half-life so that it does not need to be replaced often.

Sterilizing

Gamma rays can be used to kill bacteria, mould and insects in food even after it has been packaged.

This process prolongs the shelf-life of the food, but sometimes changes the taste.

Gamma rays are also used to sterilize hospital equipment, especially plastic syringes that would be damaged if heated.

Radioactive Dating

Animals and plants have a known proportion of carbon-14 (a radioisotope of carbon) in their tissues.

When they die they stop taking carbon in, then the amount of carbon-14 decreases at a known rate (carbon-14 has a half-life of 5700 years).

The age of ancient organic materials can be found by measuring the amount of carbon-14 that remains.

Radioactive Tracers

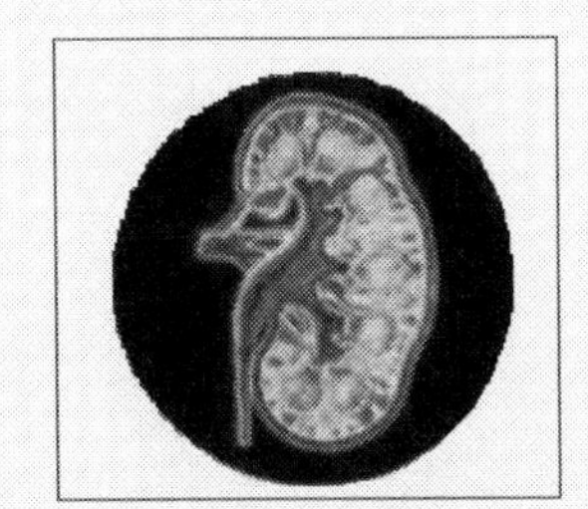

The most common tracer is technetium-99 and is very safe because it only emits gamma rays and doesn't cause much ionization.

Radioisotopes can be used for medical purposes, such as checking for a blocked kidney.

A small amount of Iodine-123 is injected into the patient, and after five minutes two Geiger counters are placed over the kidneys.

Also radioisotopes are used in industry, to detect leaking pipes. A small amount is injected into the pipe. It is then detected with a Geiger counter above ground.

Cancer Treatment

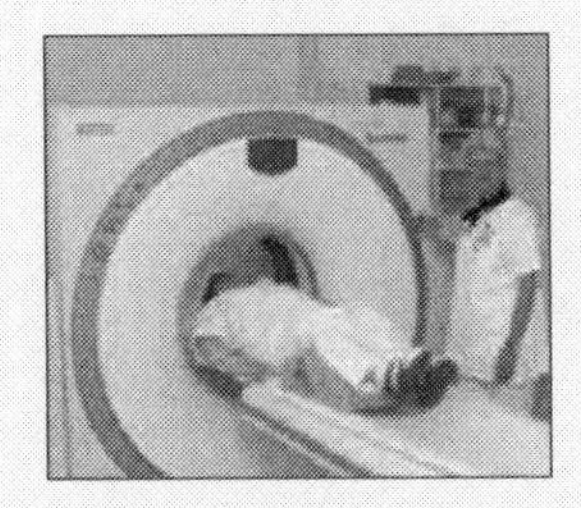

Because gamma rays can kill living cells, they are used to kill cancer cells without having to resort to difficult surgery. This is called 'radiotherapy', and works because, unlike healthy cells, cancer cells cannot repair themselves when damaged by gamma rays.

It is vital to get the dosage correct. Too great a dose will damage too many healthy cells a dose will not stop, while too small does will not stop the cancer from spreading in time.

Some cancers are easier to treat with radiotherapy than others - it's not too difficult to aim gamma rays at a breast tumor, but for lung cancer it is much harder to avoid damaging healthy cells. Also, lungs are more easily damaged by gamma rays, and therefore other treatments may be used.

Checking Welds

If a gamma source is placed on one side of the welded metal, and a photographic film on the other side, weak points or air bubbles will show up on the film, like an X-ray.

In the Petroleum Industry

Another application of radioisotopes is in the oil industry. For example, a small amount of radioisotope is placed into oil pipes in order to observe the circulation of oil. Additionally, if a single pipe is used to transfer more than one petroleum derivative (one after the other) a small amount of radioactive isotope is placed into the last portion of one substance to signal its end and the start of another.

Radiography

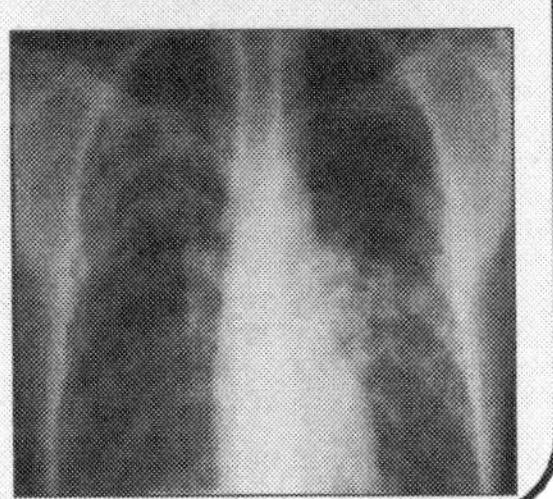

A picture can be obtained on film or sensitive plaque with the help of radioactive emissions. This is known in medicine as roentgen film. In a roentgen film X-rays are used which are produced by electronic devices.

SUPPLEMENTARY QUESTIONS

1. What are the differences and similarities between nuclear and chemical reactions? Explain.

2. When the atomic number of elements increases, the ratio between the number of neutrons (N_n) and the number of protons (N_p), (that is N_n/N_p) also increases. What is the reason for this?

3. What is a radioactive property? Which nuclei of atoms are supposed to have radioactive properties? Explain.

4. A radioactive compound $XYZ_3(s)$ is decomposed according to the reaction below:

   $XYZ_3(s) \rightarrow XY(s) + 3/2Z_2(g)$

   If the compound XY(s) is not radioactive, what can be said about radioactivity of elements X, Y and Z?

5. What are the properties of alpha emission? Which particles are ejected if a nucleus of atom emits an α particle?

6. What are the properties of beta emission?

7. What is the source of gamma (ψ) rays? Explain.

8. Although α and β^- rays deviate in the electrical and magnetic fields, ψ rays do not. Why not? Explain.

9. Of alpha, beta and gamma rays:

   a. Which has the greatest penetrating power?

   b. Which ionizes atoms mostly?

10. How does a nucleus that emits a positron particle change?

11. When a proton of a radioactive nucleus is transmuted into a neutron, which particles are produced?

12. If a nucleus captures an electron from K shell, how will the number of protons in the nucleus be changed?

13. Why are X-rays emitted when a nucleus captures an electron?

14. Find the particles which are represented by X in the following nuclear reaction equations.

    a. ${}^{218}_{84}Po \rightarrow {}^{218}_{85}At + X$ **b.** ${}^{234}_{92}U \rightarrow {}^{226}_{88}Ra + 2X$

    c. ${}^{30}_{15}P \rightarrow {}^{30}_{14}Si + X$ **d.** ${}^{9}_{4}Be + \alpha \rightarrow {}^{12}_{6}C + X$

15. Find the values of A and Z in the following nuclear reaction equations.

    a. ${}^{234}_{90}Th \rightarrow {}^{A}_{90}Th + \alpha + 2\beta^-$

    b. ${}^{23}_{12}Mg \rightarrow {}^{23}_{Z}Na + \beta^+$

    c. ${}^{14}_{7}N + n \rightarrow {}^{A}_{Z}C + p$

    d. ${}^{A}_{Z}Bi + \alpha \rightarrow {}^{211}_{85}At + 2n$

16. If an atom radiates 1α and $2\beta^-$ emissions, what will be the change in the atomic mass number and in the atomic number of the atom?

17. When the element X emits two α particles successively, the isotope ${}^{218}_{84}Po$ forms. Find the number of neutrons of X.

18. If the radioactive nucleus of ${}^{234}_{91}Pa$ emits 2α and $1\beta^-$ particles successively, what will be the number of neutrons of the new nucleus produced?

19. $X \rightarrow {}^{222}_{86}Rn + 3\alpha + 2\beta^-$

    What are the atomic and the atomic mass numbers of the element X?

20. When the radioactive element X radiates 8α and $6\beta^-$ particles successively, the element ${}^{206}_{82}Pb$ forms. According to this, what is the atomic number of the element X?

21. When the element X in the 8A group in the periodic table radiates 2α and $1\beta^-$ particles, the element Y forms. Find the group number of Y in the periodic table.

22. When the isotope ${}^{234}_{90}Th$ emits 4α and $n\beta^-$ particles, the element At forms. If the number of neutrons of At is 133, what is the value of n?

23. ${}^{238}_{92}X \xrightarrow{\alpha} Y \xrightarrow{\beta^-} Z \xrightarrow{\beta^-} T$

    The decay series started by ${}^{238}_{92}X$ and containing several emissions is represented above. According to this, which two of X, Y, Z and T are isotopes of each other? Explain.

24. ${}^{7}_{3}Li + {}^{1}_{1}p \longrightarrow {}^{7}_{4}Be + X$

    What is the particle X produced by the nuclear reaction above?

25. $^{20}_{10}Ne + ^{1}_{0}n \longrightarrow X + \alpha$

What is the number of neutrons of the element X in the above nuclear equation?

26. $X + ^{4}_{2}He \longrightarrow Y + ^{1}_{0}n$

What is the atomic number of the element X, if the element Y is in the 4th period and the 7A group in the periodic table?

27. $^{11}_{5}B + \alpha \longrightarrow X + ^{1}_{0}n$

$X + \alpha \longrightarrow ^{17}_{8}O + Y$

What is the formula of the stable compound of the elements X and Y?

28. If a nucleus of $^{27}_{13}Al$ is bombarded with neutrons, the nucleus transmutes into the nucleus of element X by capturing a neutron and emitting an alpha particle. What are the numbers of protons and neutrons in the nucleus of the element X?

29. What are the factors affecting the rate of radioactive decay? Does temperature affect the rate of radioactive decay? Explain.

30. How many hours does it take to decay 3/4 of an element which has half-life of 15 hours?

31. After 24 hours, 12.5% of a radioactive element remains undecayed. What is the half-life of the element?

32. The half-life of a radioactive substance is 3 days. How many days does it take to decay 93.75% of this substance?

33. The half-life of the radioactive element $^{35}_{16}S$ is 88 days. An amount of S decays for 440 days, after which 2 grams remains undecayed. What is the initial mass of $^{35}_{16}S$ in grams?

34. The half-life of the radioactive isotope $^{28}_{12}Mg$ is 21 hours. How many grams of 1 mol of $^{28}_{12}Mg$ remains undecayed after 63 hours?

35. After 12 days, the mass of the radioactive element X with a half-life of 4 days decreases by 21 g. What is the initial mass of the element X?

36. What is the initial mass of a radioactive element which loses 45 g after halving 4 times?

37. 75% of a radioactive element X decays in n years. How many grams of 32 g of the element decays in 2n years?

38. The isotope $^{210}_{83}Bi$ has a half-life of 5 days. If the differences between the amount of Bi after 10 days and the amount of Bi after 20 days is 75 g, what is the mass of Bi remaining after 25 days?

39. Suppose that, at 14^{00} hours, a radioactive isotope weighs $2.56 \cdot 10^{-3}$ mg. At 16^{00} hours on the same day $4 \cdot 10^{-5}$ mg of substance remains. What is the half-life of this isotope in minutes?

40. After 6 hours, 15 g of 100 g mixture of radioactive elements X and Y remain. X has a half-life of 2 hours and Y has a half-life of 3 hours. Calculate the masses of X and Y in the initial mixture in grams.

41. What is the most hazardous radiation to human beings? Explain.

42. What changes are caused by radioactive rays on the human organism?

43. What are the possible illnesses that can appear in the human being exposed to a high dose of radiation?

44. What are the radioactive isotopes and their fields of usage in medicine?

45. What are cosmic rays?

46. Which creature is the most durable in the face of radioactivity?

47. Which rays are used to control the thickness of paper?

48. Which element is used in smoke detectors?

49. Which rays are used to sterilize foods?

50. Which carbon isotope is used for radioactive dating?

51. What is the reason of using gamma rays (ψ) for the treatment of cancer?

52. Which rays are used for taking roentgen film?

MULTIPLE CHOICE QUESTIONS

1. Which of the reactions below represent(s) a radioactive decaying?

I. ${}^{23}_{11}Na$ + Energy $\longrightarrow$ ${}^{23}_{11}Na^+ + e^-$

II. ${}^{23}_{12}Mg \longrightarrow {}^{23}_{11}Na + {}^{0}_{+1}e$

III. ${}^{14}_{6}C \longrightarrow {}^{14}_{7}N + {}^{0}_{-1}e$

A) I only B) II only C) I and II D) II and III E) I, II and III

2. Compounds XY and X_2Z are radioactive, but XT is not radioactive. Which one of the substances given below does not have radioactive properties?

A) ZY_2 B) YT C) X_2 D) Z_2 E) T_2Z

3. Which of the forms of radiation below changes the nucleus of the element?

I. Alpha particle emission

II. Beta particle emission

III. Gamma radiation

A) I only B) II only C) I and II D) II and III E) I, II and III

4. Which of the statement(s) below is/are true?

I. Alpha radiations are 2+ charged

II. Beta particles are 1– charged electrons

III. If an atom emits a positron particle, its atomic number increases.

A) I only B) II only C) I and II D) II and III E) I, II and III

5. Which of the statement(s) below will be true if a neutron is transformed into a proton in the nucleus of an atom?

I. A positron particle is ejected from nucleus.

II. The atomic number is increased by 1.

III. A beta particle is emitted.

A) I only B) II only C) I and II D) II and III E) I and III

6. A radioactive element X emits one alpha, one beta and one neutron particles. After this reaction, how does the atomic number of X change?

A) decreases by 1 B) decreases by 2 C) decreases by 3 D) increases by 1 E) no change

7. Which properties below is/are changed if a proton is converted into a neutron in the nucleus of a radioactive atom X?

I. The atomic radius of X.

II. The placement of X in the periodic table.

III. The chemical properties of X.

A) II only B) I and II C) I and III D) II and III E) I, II and III

8. Which of the statement(s) below will be true if a nucleus of a radioactive atom captures one neutron:

I. Its isotope forms.

II. Its chemical property changes.

III. Its atomic mass number increases.

A) I only B) II only C) I and II D) I and III E) I, II and III

9. What is the number of neutrons of X in the nuclear equation below?

$X \longrightarrow {}^{218}_{85}At + 4\alpha + 1\beta^-$

A) 140 B) 141 C) 142 D) 143 E) 144

10. When a radioactive isotope ${}^{230}_{90}Th$ emits 3α and $1\beta^-$ which one of the following atoms is formed?

A) ${}^{218}_{84}Po$ B) ${}^{218}_{85}At$ C) ${}^{214}_{84}Po$ D) ${}^{220}_{86}Rn$ E) ${}^{216}_{85}At$

11. Which of the statement(s) below is/are true when a radioactive element X in the 8A group transmutes into element Y by emitting 1α and $1\beta^-$ particles. So;

I. X and Y are isotopes with each other.

II. The number of neutrons of X is more than that of Y by 3.

III. Y is a halogen.

A) I only B II only C) I and II D) I and III E) II and III

12. 0.2 mol of $^{234}_{91}Pa$ is disintegrated into exactly 0.2 mol of β^- particles and 8.96 L He(g) at STP are produced. What is the product of this nuclear reaction?

A) $^{234}_{92}U$ B) $^{234}_{90}Th$ C) $^{226}_{88}Ra$ D) $^{226}_{89}Ac$ E) $^{228}_{88}Ra$

13. Which of the statement(s) below will be incorrect if the bombardment of atom X by one α particle produces a nucleus of $^{12}_{6}C$ and a neutron?

I. The atomic number of X is 4.

II. The number of neutrons in X is 4.

III. X is an isotope of $^{8}_{4}Be$.

A) I only B) II only C) I and II D) II and III E) I and III

14. The bombardment of by 1α particle atom X, which is a member of the 5A group in the periodic table, produces a nucleus Y and one neutron. What is the group number of Y?

A) 4A B) 5A C) 6A D) 7A E) 8A

15. In the nuclear reaction equation below, element X is produced. What is the formula of the compound formed between element X and oxygen? ($_8O$)

$$^{27}_{13}Al + ^{1}_{0}n \longrightarrow X + \alpha$$

A) XO B) X_2O C) XO_2 D) X_2O_3 E) X_2O_5

16. The bombardment of atom X by a $^{1}_{0}n$ produces $^{24}_{11}Na$ and 1α. What is the molecular mass of the compound which is formed between X and sulfur? ($^{32}_{16}S$)

A) 59 B) 74 C) 118 D) 145 E) 150

17. What is X in the nuclear fission reaction below?

$$^{235}_{92}U + ^{1}_{0}n \longrightarrow ^{90}_{38}Sr + ^{144}_{54}Xe + 2X$$

A) $^{1}_{1}H$ B) $^{4}_{2}He$ C) $^{1}_{0}n$ D) $^{0}_{-1}e$ E) $^{0}_{+1}e$

18. Which one of the following factors has an effect on the half-life of a radioactive element?

A) Temperature
B) Pressure
C) Solubility in water
D) Type of element
E) Amount of element

19. The half-life of $^{218}_{84}Po$ is 3 minutes. What amount of a 1 mol sample of $^{218}_{84}Po$ in grams remains after 9 minutes?

A) 13.625 B) 27.25 C) 109 D) 166 E) 218

20. The half-life of $^{210}_{83}Bi$ is 5 days. How long does it take to decay 15/16 of $^{210}_{83}Bi$?

A) 10 B) 15 C) 20 D) 25 E) 30

21. The difference in the mass of a radioactive element X between the 2nd and 3rd decays is 90 g. What is the mass of undecayed X after the 4th decay in grams?

A) 180 B) 90 C) 45 D) 30 E) 15

22. Which gives the correct order according to durability in the face of radioactivity?

A) Snail > Dog > Scorpion > Rabbit > Human
B) Scorpion > Snail > Rabbit > Human > Dog
C) Scorpion > Snail > Human > Rabbit > Dog
D) Snail > Scorpion > Human > Rabbit > Dog
E) Snail > Dog > Rabbit > Human > Scorpion

23. Which one of the following is not a result of radioactivity ?

A) Cancer
B) DNA structure change
C) Mutation
D) Tumors
E) Tuberculosis

24. Which is not a usage of radioactivity?

A) Thickness control of paper
B) Radioactive dating
C) Sterilizing of foods
D) Cancer treatment
E) Welding

25. Respectively, which rays are used for sterilizing foods, detecting smoke, controlling the thickness of paper and checking welds?

A) $\alpha, \beta^-, \psi, \psi$ B) $\psi, \alpha, \beta^-, \psi$ C) $\psi, \psi, \alpha, \beta^+$ D) $\psi, \beta^+, \alpha, \beta^-$ E) $\beta^-, \beta^+, \alpha, \psi$

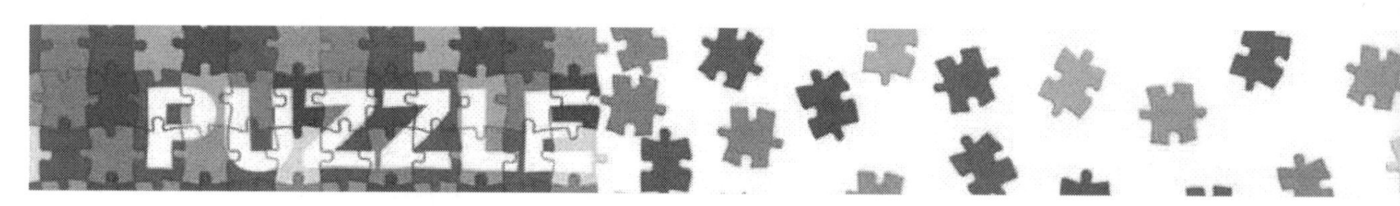

CROSSWORD PUZZLE

ACROSS

1 Type of electromagnetic ray that is stopped by 5 sheets of aluminum foil.
5 Type of electromagnetic ray with a very high energy stopped by 3 cm steel block.
8 Process in which very light nuclei are fused into heavier nuclei.
9 Type of electromagnetic ray with low penetrating power stopped by a sheet of paper.
11 Splitting of a nucleus into smaller nuclei.
12 Name of atomic bomb dropped on Hiroshima.
13 Scientist who discovered alpha rays.

DOWN

1 French scientist who discovered radioactivity.
2 Period of time required for half of the initial amount of a substance to disintegrate.
3 Name of atomic bomb dropped on Nagasaki.
4 Spontaneous emission of radioactive rays by an unstable atomic nucleus.
6 Positively charged nuclear particle with the same mass as an electron.
7 Scientist who achieved the first fission reaction.
10 German scientist who first discovered X-rays.

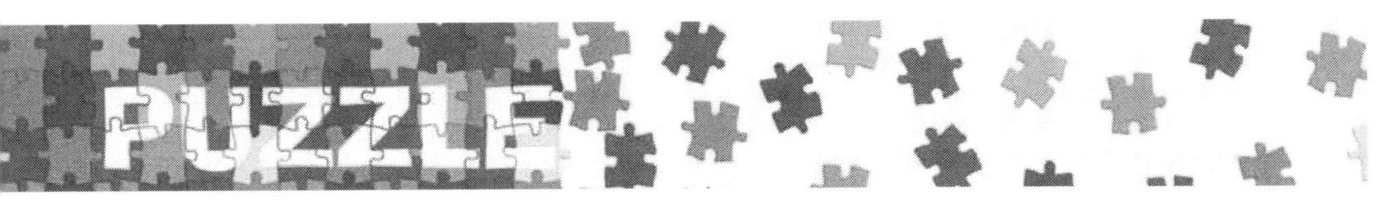

WORD SEARCH PUZZLE

Find the words in the grid. Words can go horizontally, vertically and diagonally in all eight directions.

M	Y	N	B	Q	M	A	R	I	I	C	▪	R	I	I	L	V	C	N
H	L	Y	O	Y	G	N	I	T	A	D	N	O	B	R	A	C	N	N
K	H	G	A	I	N	€	M	T	N	X	N	T	H	M	N	T	N	O
€	N	C	L	L	S	B	Q	B	R	I	M	I	S	S	I	O	N	R
W	I	R	V	T	M	▪	D	M	G	L	H	Q	X	W	D	R	P	T
D	N	O	I	S	S	I	F	T	K	K	U	€	R	L	B	V	R	I
Q	C	C	Q	C	V	R	N	R	N	Y	Y	P	I	Y	T	L	I	S
N	Q	B	N	U	H	I	R	G	I	H	N	U	V	€	P	N	L	O
▪	R	I	M	B	O	I	A	A	P	K	T	G	I	F	R	P	L	P
C	▪	T	U	R	I	M	R	A	D	G	A	H	T	I	X	F	▪	C
L	T	A	K	Q	M	C	R	▪	H	I	A	S	C	N	T	B	M	H
I	H	M	F	A	L	G	Q	R	T	L	A	O	A	V	L	N	R	I
A	I	R	Y	F	O	Y	K	▪	F	P	F	T	O	G	F	P	I	R
R	R	U	T	I	N	L	N	L	I	I	A	T	I	M	A	U	G	O
T	F	C	D	B	G	F	I	A	R	R	W	C	D	O	T	N	I	S
H	O	A	N	R	K	F	B	M	L	T	I	N	A	R	N	K	I	H
C	R	K	X	P	I	M	I	D	N	P	G	L	R	P	K	P	G	I
P	D	K	B	V	R	H	L	Y	V	H	H	M	F	R	Y	A	R	M
G	Y	T	O	M	O	G	R	A	P	H	Y	A	Q	V	T	M	H	A

ALPHA	DECAY	GAMMA	NAGASAKI	RADIOACTIVE
BECQUEREL	EMISSION	GEIGERMULLER	NUCLEAR	RAY
BETA	ENRICOFERMI	HALFLIFE	POSITRON	ROENTGEN
CAPTURE	FISSION	HIROSHIMA	RADIOGRAPHY	RUTHERFORD
CARBONDATING	FUSION	MARIECURIE	RADIATION	TOMOGRAPHY

NOTES

APPENDICES
GLOSSARY
ANSWERS
INDEX

APPENDICES

Appendix A

Electron Configurations of the Elements

1. H : $1s^{1}$
2. He : $1s^{2}$
3. Li : [He] $1s^{1}$
4. Be : [He] $2s^{2}$
5. B : [He] $2s^{2}2p^{1}$
6. C : [He] $2s^{2}2p^{2}$
7. N : [He] $2s^{2}2p^{3}$
8. O : [He] $2s^{2}2p^{4}$
9. F : [He] $2s^{2}2p^{5}$
10. Ne : [He] $2s^{2}2p^{6}$
11. Na : [Ne] $3s^{1}$
12. Mg : [Ne] $3s^{2}$
13. Al : [Ne] $3s^{2}3p^{1}$
14. Si : [Ne] $3s^{2}3p^{2}$
15. P : [Ne] $3s^{2}3p^{3}$
16. S : [Ne] $3s^{2}3p^{4}$
17. Cl : [Ne] $3s^{2}3p^{5}$
18. Ar : [Ne] $3s^{2}3p^{6}$
19. K : [Ar] $4s^{1}$
20. Ca : [Ar] $4s^{2}$
21. Sc : [Ar] $4s^{2}3d^{1}$
22. Ti : [Ar] $4s^{2}3d^{2}$
23. V : [Ar] $4s^{2}3d^{3}$
24. Cr : [Ar] $4s^{1}3d^{5}$
25. Mn : [Ar] $4s^{2}3d^{5}$
26. Fe : [Ar] $4s^{2}3d^{6}$
27. Co : [Ar] $4s^{2}3d^{7}$
28. Ni : [Ar] $4s^{2}3d^{8}$
29. Cu : [Ar] $4s^{1}3d^{10}$
30. Zn : [Ar] $4s^{2}3d^{10}$
31. Ga : [Ar] $4s^{2}3d^{10}4p^{1}$
32. Ge : [Ar] $4s^{2}3d^{10}4p^{2}$
33. As : [Ar] $4s^{2}3d^{10}4p^{3}$
34. Se : [Ar] $4s^{2}3d^{10}4p^{4}$
35. Br : [Ar] $4s^{2}3d^{10}4p^{5}$
36. Kr : [Ar] $4s^{2}3d^{10}4p^{6}$
37. Rb : [Kr] $5s^{1}$
38. Sr : [Kr] $5s^{2}$
39. Y : [Kr] $5s^{2}4d^{1}$
40. Zr : [Kr] $5s^{2}4d^{2}$
41. Nb : [Kr] $5s^{1}4d^{4}$
42. Mo : [Kr] $5s^{1}4d^{5}$
43. Tc : [Kr] $5s^{1}4d^{6}$
44. Ru : [Kr] $5s^{1}4d^{7}$
45. Rh : [Kr] $5s^{1}4d^{8}$
46. Pd : [Kr] $4d^{10}$
47. Ag : [Kr] $5s^{1}4d^{10}$
48. Cd : [Kr] $5s^{2}4d^{10}$
49. In : [Kr] $5s^{2}4d^{10}5p^{1}$
50. Sn : [Kr] $5s^{2}4d^{10}5p^{2}$
51. Sb : [Kr] $5s^{2}4d^{10}5p^{3}$
52. Te : [Kr] $5s^{2}4d^{10}5p^{4}$
53. I : [Kr] $5s^{2}4d^{10}5p^{5}$
54. Xe : [Kr] $5s^{2}4d^{10}5p^{6}$
55. Cs : [Xe] $6s^{1}$
56. Ba : [Xe] $6s^{2}$
57. La : [Xe] $6s^{2}5d^{1}$
58. Ce : [Xe] $6s^{2}4f^{1}5d^{1}$
59. Pr : [Xe] $6s^{2}4f^{3}$
60. Nd : [Xe] $6s^{2}4f^{4}$
61. Pm : [Xe] $6s^{2}4f^{5}$
62. Sm : [Xe] $6s^{2}4f^{6}$
63. Eu : [Xe] $6s^{2}4f^{7}$
64. Gd : [Xe] $6s^{2}4f^{7}5d^{1}$
65. Tb : [Xe] $6s^{2}4f^{9}$
66. Dy : [Xe] $6s^{2}4f^{10}$
67. Ho : [Xe] $6s^{2}4f^{11}$
68. Er : [Xe] $6s^{2}4f^{12}$
69. Tm : [Xe] $6s^{2}4f^{13}$
70. Yb : [Xe] $6s^{2}4f^{14}$
71. Lu : [Xe] $6s^{2}4f^{14}5d^{1}$
72. Hf : [Xe] $6s^{2}4f^{14}5d^{2}$
73. Ta : [Xe] $6s^{2}4f^{14}5d^{3}$
74. W : [Xe] $6s^{2}4f^{14}5d^{4}$
75. Re : [Xe] $6s^{2}4f^{14}5d^{5}$
76. Os : [Xe] $6s^{2}4f^{14}5d^{6}$
77. Ir : [Xe] $6s^{2}4f^{14}5d^{7}$
78. Pt : [Xe] $6s^{1}4f^{14}5d^{9}$
79. Au : [Xe] $6s^{1}4f^{14}5d^{10}$
80. Hg : [Xe] $6s^{2}4f^{14}5d^{10}$
81. Tl : [Xe] $6s^{2}4f^{14}5d^{10}6p^{1}$
82. Pb : [Xe] $6s^{2}4f^{14}5d^{10}6p^{2}$
83. Bi : [Xe] $6s^{2}4f^{14}5d^{10}6p^{3}$
84. Po : [Xe] $6s^{2}4f^{14}5d^{10}6p^{4}$
85. At : [Xe] $6s^{2}4f^{14}5d^{10}6p^{5}$
86. Rn : [Xe] $6s^{2}4f^{14}5d^{10}6p^{6}$
87. Fr : [Rn] $7s^{1}$
88. Ra : [Rn] $7s^{2}$
89. Ac : [Rn] $7s^{2}6d^{1}$
90. Th : [Rn] $7s^{2}6d^{2}$
91. Pa : [Rn] $7s^{2}5f^{2}6d^{1}$
92. U : [Rn] $7s^{2}5f^{3}6d^{1}$
93. Np : [Rn] $7s^{2}5f^{4}6d^{1}$
94. Pu : [Rn] $7s^{2}5f^{6}$
95. Am: [Rn] $7s^{2}5f^{7}$
96. Cm: [Rn] $7s^{2}5f^{7}6d^{1}$
97. Bk : [Rn] $7s^{2}5f^{9}$
98. Cf : [Rn] $7s^{2}5f^{10}$
99. Es : [Rn] $7s^{2}5f^{11}$
100. Fm : [Rn] $7s^{2}5f^{12}$
101. Md : [Rn] $7s^{2}5f^{13}$
102. No : [Rn] $7s^{2}5f^{14}$
103. Lr : [Rn] $7s^{2}5f^{14}6d^{1}$
104. Rf : [Rn] $7s^{2}5f^{14}6d^{2}$
105. Db : [Rn] $7s^{2}5f^{14}6d^{3}$
106. Sg : [Rn] $7s^{2}5f^{14}6d^{4}$
107. Bh : [Rn] $7s^{2}5f^{14}6d^{5}$
108. Hs : [Rn] $7s^{2}5f^{14}6d^{6}$
109. Mt : [Rn] $7s^{2}5f^{14}6d^{7}$
110. Ds : [Rn] $7s^{1}5f^{14}6d^{9}$
111. Rg : [Rn] $7s^{1}5f^{14}6d^{10}$
112. Uub : [Rn] $7s^{2}5f^{14}6d^{10}$
114. Uuq : [Rn] $7s^{2}5f^{14}6d^{10}7p^{2}$
116. Uuh : [Rn] $7s^{2}5f^{14}6d^{10}7p^{4}$

Appendix B

The letters s, p, d and f originated from the words 'sharp', 'principal', 'diffuse' and 'fundamental', which are used to define the atomic spectra of the alkali metals. However, starting with the letter f, the orbital designations follows alphabetical order.

Quantum Numbers

The positions of electrons around the nucleus are determined with the help of four quantum numbers. There is the **principal quantum number (n)**, **secondary quantum number (l)**, **magnetic quantum number (m_l)** and **spin quantum number (m_s)**. Two electrons in an atom never have identical sets of the four quantum numbers. At least one of the four quantum numbers must be different. This is known as **Pauli's principle**.

1. The Principal Quantum Number (n)

Principal quantum numbers are related to the distance of an electron cloud from the nucleus. They represent energy levels, which are also called **shells**. The principal quantum number is denoted by n. The value of n is a positive integer such as 1,2,3,4... and each shell can be also denoted by the letters. K, L, M, N...

Principal quantum number (n) : 1,2,3,4...

Shell notation by letters : K,L,M,N...

The principal quantum number, n, corresponds to the energy levels in Bohr's theory. The bigger the value of n, the further the electron cloud from the nucleus, hence the higher the potential energy of the electron.

2. The Secondary (Orbital) Quantum Number (l)

In his uncertainty principle, Heisenberg expressed the fact that the orbitals have to be considered as electron clouds rather than the definite circular paths of electron orbit around the nucleus. These electron clouds form an electric field in the atom. There can be some divisions in these energy levels according to the effects of external electric fields (created by other electrons or external sources). As a result, each energy level has sub-energy shells.

Since the energy levels have different shapes, secondary quantum numbers are used in order to represent the shape of the electron clouds and divisions of the energy levels. **The secondary quantum number** is also called the orbital quantum number and denoted by l. The values of the secondary quantum number are the positive numbers from zero to n–1, depending on the principal quantum number. The secondary quantum numbers are also represented by letters such as s,p,d and f... starting from the least energetic.

Secondary quantum numbers (l) : 0,1,2,3...

Subshell notations : s,p,d,f...

The energies of subshells in the same shell (energy level) increase in the order $s<p<d<f$.

Now, let us examine the subshells s, p, d and f in detail.

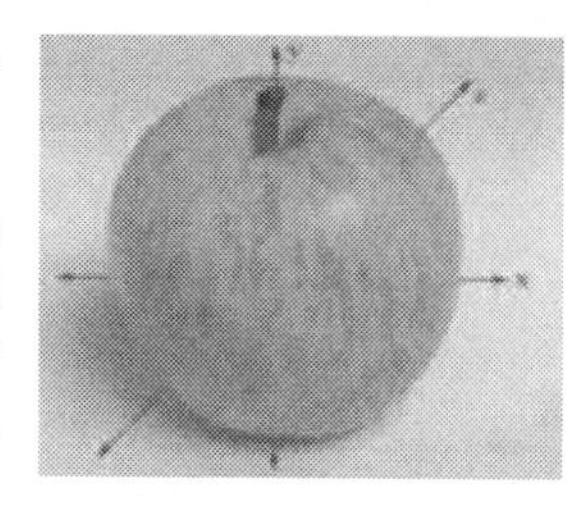

The s Subshell

All orbitals with a value of l = 0 are orbitals of the s subshell. If the s subshell is in the first energy level (n=1) it is denoted by 1s, if it is in the second energy level (n=2) it is denoted by 2s, in the third shell by 3s, and so on.

Orientation of an s orbital in space

The electron clouds of the s orbital (or probability distribution of finding the electron) has spherical symmetry. The s orbital is the shape of a sphere, in which the electron cloud density becomes less dense from the center outward, and the nucleus is at the center. Look at the figure above right.

The p Subshell

All orbitals with a value of l = 1, are the orbitals of the p subshell. The p orbitals are not spherical. Each p subshell consists of three orbitals in the form of lobes that differ in their orientation. These lobes are separated from each other by a plane where the probability of finding the electron is zero. The lobes are located on both sides of this plane like a dumbbell. The shape of the p orbitals are the same but the directions of the lobes are different. Since it is possible to imagine that these lobes are oriented along x, y and z coordinates, so the corresponding p orbitals are denoted by p_x, p_y and p_z. Hence, in all main energy levels, except first energy level (n = 1), there are three p orbitals.

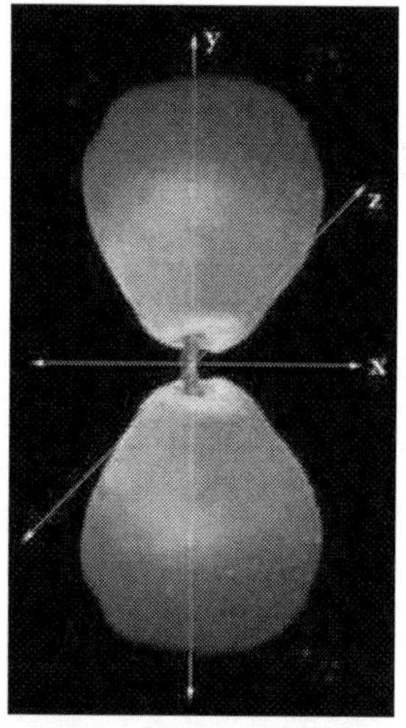

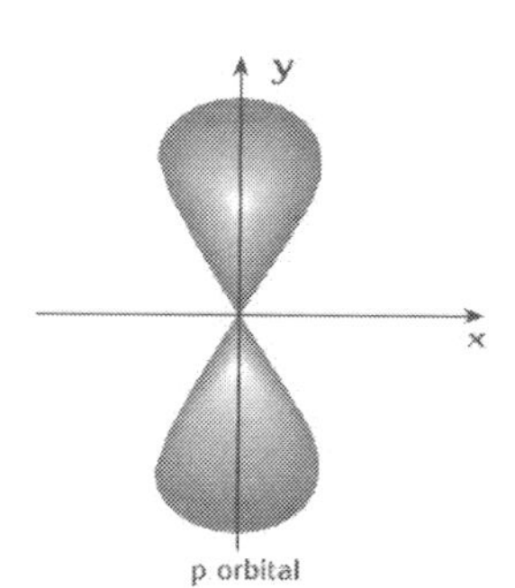

The two lobes of a p orbital have different special orientations.

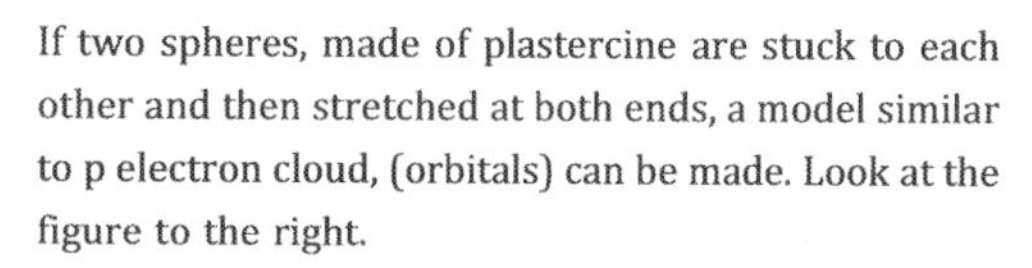

If two spheres, made of plastercine are stuck to each other and then stretched at both ends, a model similar to p electron cloud, (orbitals) can be made. Look at the figure to the right.

The d Subshell

All orbitals with a value of l = 2 are orbitals of the d subshell. The geometry of d orbitals is more complex than the s and p orbitals, and each d subshell has five orbitals with specific orientations. Look at the figure to the right.

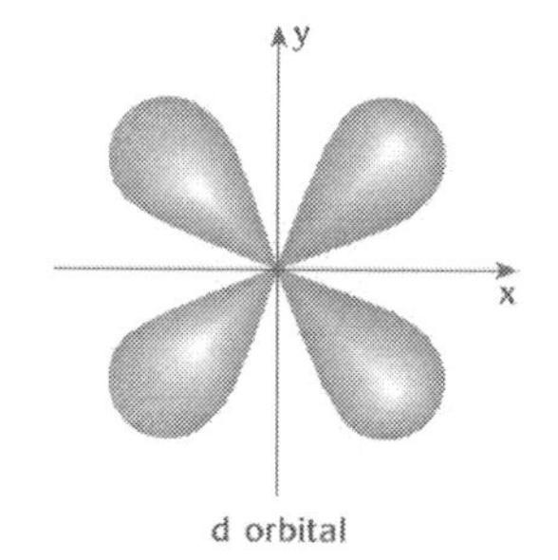

Schematic representation of any d orbital.

The f Subshell

All orbitals with a value of l = 3 are orbitals of the f subshell. The shape of the f orbitals is much more complex than the shape of the s, p and d orbitals. Each subshell has seven orbitals, each with specific orientations.

The subshells and the types of orbitals for the first four energy levels (principle quantum numbers) are shown below.

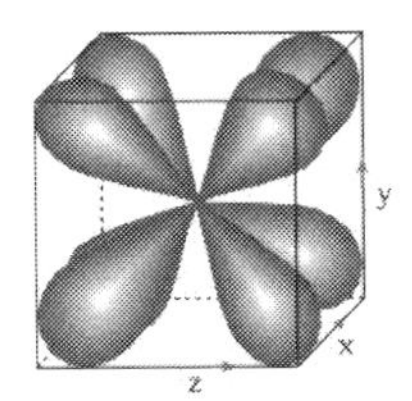

f orbital

Schematic representation of any f orbital.

n	1	2		3			4			
l	0	0	1	0	1	2	0	1	2	3
Orbital	1s	2s	2p	3s	3p	3d	4s	4p	4d	4f

The orbitals and the subshell representation for the first four energy levels.

Even if the shapes of the orbitals for the first four subshells are given, the shapes of the d and f subshell orbitals are so detailed at this point that they are given to students who are especially interested in learning more about orbital geometry. Look at the figure to the left.

3. The Magnetic Quantum Number (m_l)

Each subshell consists of one or more orbitals. Each orbital in each subshell is denoted by a **magnetic quantum number, m_l**. This number is deduced as a result of the observation of the new lines in the atomic spectrum in a magnetic field. The magnetic quantum number has a value which ranges from –l to +l.

$m_l = (-l), -(l-1), -(l-2) \dots 0 \dots (l-2), (l-1), (l)$

The number of the secondary quantum number in a given shell is equal to the value of the principal quantum number of that shell. The number of orbitals in a given subshell is calculated by the equation of **$2l + 1$**.

Now, let us examine the orbitals in some shells.

For n = 1, the secondary quantum number (l) takes only the value of 0 (zero). Thus the number of orbitals in this subshell is

$m_l = 2l + 1 = 2 \cdot 0 + 1 = 1$

This means that there is only one orbital, which is the s orbital in the first shell. Since l =0, the only acceptable m_l value for the s orbital subshell is also 0 (zero), (m_l= 0).

We can express the representation of subshells by just writing the value of the principal quantum number together with the subshell notation, so for n = 1

Principal Quantum Number(n)		Secondary Quantum Number(l)		Magnetic Quantum Number(m_l)
1	K	0	s	0
2	L	0	s	0
		1	p	-1 0 +1
3	M	0	s	0
		1	p	-1 0 +1
		2	d	+2 -1 0 +1 +2
4	N	0	s	0
		1	p	-1 0 +1
		2	d	-2 -1 0 +1 +2
		3	f	-3 -2 -1 0 +1 +2 +3

The secondary and magnetic quantum numbers for the first four energy levels.

the corresponding subshell representation is 1s where 1 is the number of the shell (n), and s is the subshell notation.

For n = 2, the secondary quantum number, l, has the values of 0 and 1. Since there are two possible secondary quantum numbers, there are two subshells as well. For l=0 the corresponding subshell is s, and for l =1, the subshell is p. Let us find the number of orbitals in these subshells.

For $l = 0$, $m_l = (2l + 1) = (2 \cdot 0 + 1) = 1$

That means, since for l=0 the only allowed value for m_l in the s subshell is 0, there is one orbital corresponding to this value of m_l, which is the 2s orbital.

For $l = 1$, $(2l + 1) = (2 \cdot 1 + 1) = 3$

This means that since for $l = 1$, m_l takes the values from -1 to +1 including zero, then there are three orbitals corresponding to these values of m_l, which are the 2p orbitals. These orbitals are denoted by p_x, p_y and p_z corresponding to the values of m_l −1, 0 and +1 respectively.

There are a total of four orbitals in second shell: one orbital from 2s and three orbitals from 2p.

4. The Magnetic Spin Quantum Number (m_s)

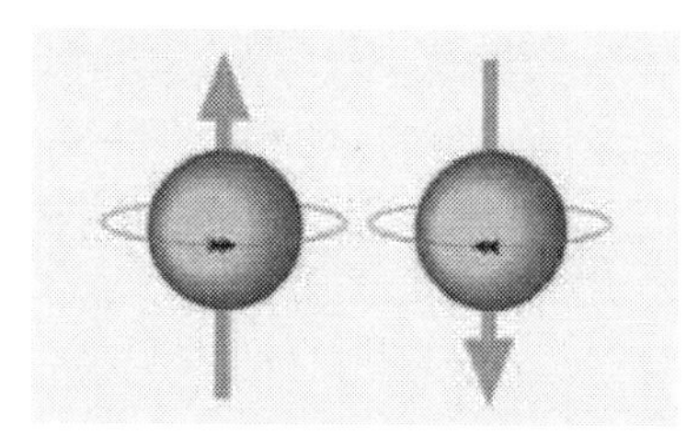

Electron spin (m_s). An electron turns around an axis passing through its center. There are two possible spinning directions for an electron (clock and counter-clockwise). Thus, for the magnetic spin quantum number, there are also two possibilities;

$m_s = +\frac{1}{2}$ *clockwise*

$m_s = -\frac{1}{2}$ *counter-clockwise*

Magnetic spin quantum number, m_s describes the direction of the rotation of the electron around its own axis and it is denoted by m_s. Since there are two possible opposite spin directions for an electron, the values of the magnetic spin quantum number, m_s, can be +1/2 and −1/2.

Since a rotating charged particle produces magnetic field, each electron has its own magnetic momentum as a result of its rotation.

The magnetic moments of the two electrons, rotating in the opposite directions, cancel each other out. That is why an orbital can hold at most two electrons with opposite spins. The number of orbitals in a shell is directly proportional to n^2. Since one orbital can hold no more than two electrons, the maximum number of electrons in a shell can be calculated by $2n^2$.

Now, let us calculate the possible number of electrons in different shells.

For n = 1, there are $2n^2 = 2 \cdot 1^2 = 2$ electrons in the first shell. These electrons belong to the 1s orbital, and this orbital is denoted by $1s^2$ where the superscript 2, shows the number of the electrons in the orbital.
For n = 2, there are $2n^2 = 2 \cdot 2^2 = 8$ electrons in the second shell. Two of them belong to the 2s orbital and six of them belong to the 2p orbitals. They are represented by $2s^2\ 2p^6$.

The Orientations of s, p, d and f Orbitals

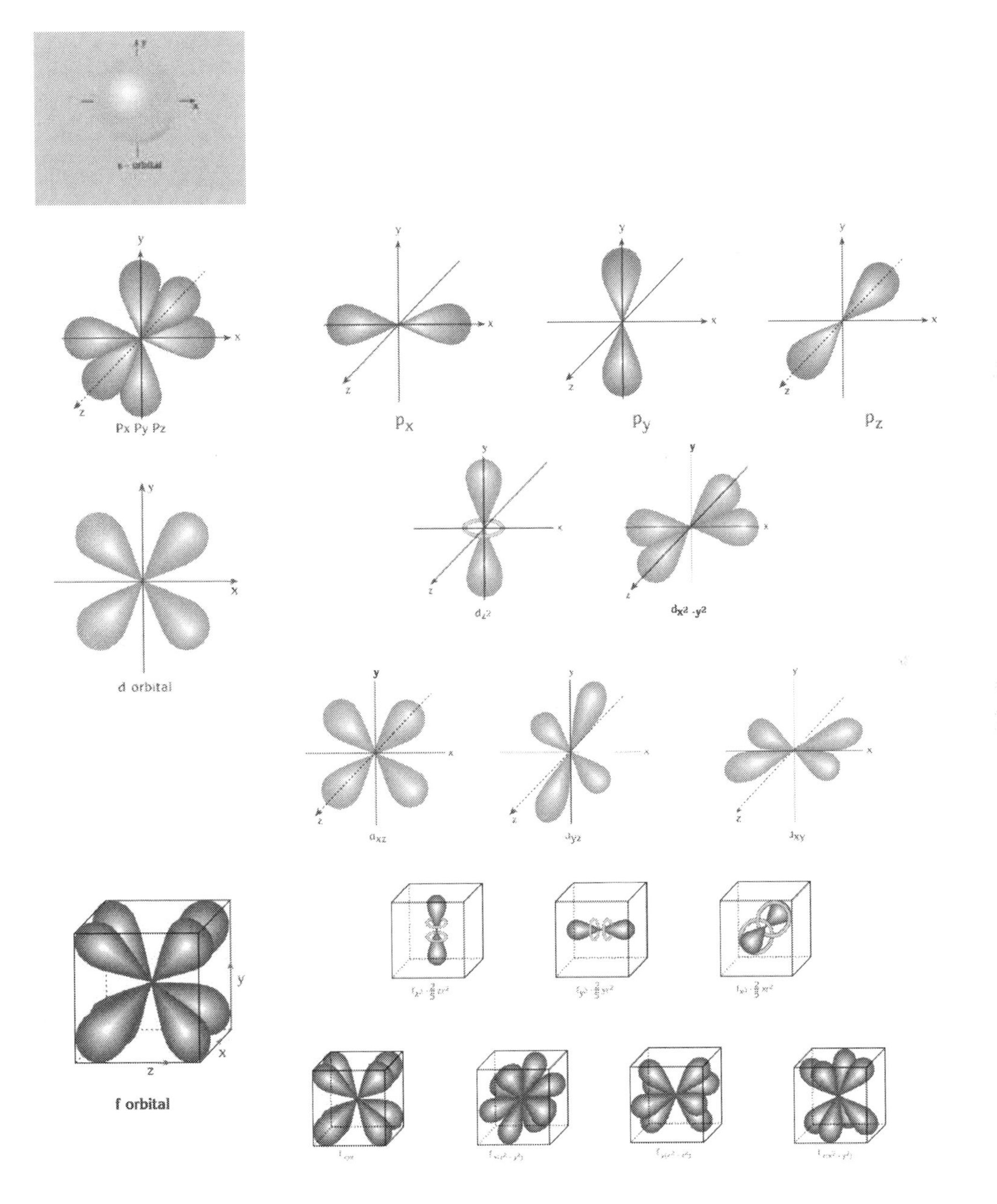

Appendix C

Discovery Date	Name	Sym	Z	Discoverer	Discovery Location	Meaning and origin of element names
Periodic Table of Elements — This list includes the 114 officially named elements.						
Archaic	Carbon	C	6	known to the ancients	Unknown	Latin word "carbo" meaning "charcoal"
Archaic	Sulfur	S	16	known to the ancients	Unknown	Sanskrit word "sulvere" meaning "sulfur";
Archaic	Iron	Fe	26	known to the ancients	Unknown	Latin word "ferrum"
Archaic	Copper	Cu	29	known to the ancients	Unknown	Latin word "cuprum" meaning the island of "Cyprus"
Archaic	Silver	Ag	47	known to the ancients	Unknown	Latin word "argentum" meaning "silver"
Archaic	Tin	Sn	50	known to the ancients	Unknown	Latin word "stannum" meaning "tin"
Archaic	Antimony	Sb	51	known to the ancients	Unknown	Greek words "anti + monos" meaning "not alone"
Archaic	Gold	Au	79	known to the ancients	Unknown	Word "aurum" meaning "gold" in Latin.
Archaic	Mercury	Hg	80	known to the ancients	Unknown	The planet "Mercury"
Archaic	Lead	Pb	82	known to the ancients	Unknown	Latin word "plumbum" meaning "liquid silver"
Archaic	Bismuth	Bi	83	known to the ancients	Unknown	From German word "bisemutum"
1250	Arsenic	As	33	Alberts Magna	Unknown	Greek word "arsenikon" meaning "yellow orpiment"
1669	Phosphorus	P	15	Hennig Brand	Germany	Greek word "phosphoros" meaning "bringer of light"
1735	Platinum	Pt	78	Julius Scaliger	Italy	Spanish word "platina" meaning "silver"
1737	Cobalt	Co	27	George Brandt	Sweden	German word "kobald"meaning "goblin"
1746	Zinc	Zn	30	Andreas Marggraf	Germany	From the German word "zink"
1751	Nickel	Ni	28	Axel Cronstedt	Sweden	German word "kupfernickel" means Devil's copper
1766	Hydrogen	H	1	Henry Cavendish	England	Greek words "hydro genes" mean "water generator"
1772	Nitrogen	N	7	Daniel Rutherford	Scotland/Sweden	Potassium nitrate former in Greek
1774	Oxygen	O	8	J. Priestley, Carl W.Scheele	England/Sweden	Greek words "oxy genes" meaning acid former
1774	Chlorine	Cl	17	Carl Wilhelm Scheele	Sweden	Greek word "chloros" meaning "pale green"
1774	Manganese	Mn	25	Johann Gahn	Sweden	Latin word "magnes" meaning "magnet"
1778	Molybdenum	Mo	42	Carl Wilhelm Scheele	Sweden	Greek word "molybdos" meaning "lead"
1782	Tellurium	Te	52	F.Müller von Reichenstein	Romania	Latin word "tellus" meaning "earth"
1783	Tungsten	W	74	F.and J. José de Elhuyar	Spain	Swedish words "tung sten" meaning "heavy stone"
1789	Zirconium	Zr	40	Martin Klaproth	Germany	Arabic word "zargun" meaning "gold colour"
1789	Uranium	U	92	Martin Klaproth	Germany	From the planet Uranus
1790	Strontium	Sr	38	A. Crawford	Scotland	Village of "Strontian" in Scotland
1791	Titanium	Ti	22	William Gregor	England	"Titans" in Greek
1794	Yttrium	Y	39	Johann Gadolin	Finland	Village of "Ytterby" near Vaxholm in Sweden
1797	Chromium	Cr	24	Louis Vauquelin	France	Greek word "chroma" meaning "color"
1798	Beryllium	Be	4	Fredrich Wohler,A.A.Bussy	Germany/France	Greek word "beryllos" meaning "beryl"
1801	Niobium	Nb	41	Charles Hatchet	England	Greek word "Niobe" meaning "daughter of Tantalus"
1802	Tantalum	Ta	73	Anders Ekeberg	Sweden	Greek word "Tantalos" meaning "father of Niobe"
1803	Rhodium	Rh	45	William Wollaston	England	Greek word "rhodon" meaning "rose"
1803	Palladium	Pd	46	William Wollaston	England	Asteroid "Pallas" in Greek
1803	Cerium	Ce	58	W.V.Hisinger,J.Berzelius...	Sweden/Germany	From the asteroid Ceres
1804	Iodine	I	53	Bernard Courtois	France	Greek word "iodes" meaning "violet"
1804	Osmium	Os	76	Smithson Tenant	England	Greek word "osme" meaning "smell"
1804	Iridium	Ir	77	S.Tenant,A.F.Fourcory...	England/France	Greek word "iris" meaning "rainbow"
1807	Sodium	Na	11	Sir Humphrey Davy	England	From "soda"
1807	Potassium	K	19	Sir Humphrey Davy	England	Arabic word "qali" meaning alkali
1808	Magnesium	Mg	12	Sir Humphrey Davy	England	"Magnesia",place in Greece
1808	Calcium	Ca	20	Sir Humphrey Davy	England	Latin word "calx" meaning "lime"
1808	Barium	Ba	56	Sir Humphrey Davy	England	Greek word "barys" meaning "heavy"
1817	Lithium	Li	3	Johann Arfvedson	Sweden	Greek word "lithos" meaning "stone"
1817	Selenium	Se	34	Jöns Berzelius	Sweden	Greek word "selene" meaning "moon"
1817	Cadmium	Cd	48	Fredrich Stromeyer	Germany	Latin word "cadmia" meaning "calamine"
1823	Silicon	Si	14	Jöns Berzelius	Sweden	Latin word "silicis" meaning "flint"
1825	Aluminum	Al	13	Hans Christian Oersted	Denmark	Latin word "alumen" meaning "alum"
1826	Bromine	Br	35	Antoine J. Balard	France	Greek word "kryptos" meaning "hidden"
1828	Boron	B	5	H. Day,Gay-Lussac...	England/France	From Arabic word "buraq"
1828	Thorium	Th	90	Jons J. Berzelius	Sweden	From "Thor", Scandinavian mythological term
1830	Vanadium	V	23	Nils Sefström	Sweden	From "Vanadis", the meaning with beauty
1839	Lanthanum	La	57	Carl Mosander	Sweden	Greek word "lanthanein" meaning "to lie hidden"

1843	Terbium	Tb	65	Carl Mosander	Sweden	"Ytterby", a town in Sweden
1843	Erbium	Er	68	Carl Mosander	Sweden	"Ytterby", a village in Sweden
1844	Ruthenium	Ru	44	Karl Klaus	Russia	Latin word "Ruthenia" meaning "Russia"
1860	Cesium	Cs	55	G.Kirchoff, Robert Bunsen	Germany	Latin word "caesius" meaning "sky blue"
1861	Rubidium	Rb	37	R. Bunsen, G. Kirchoff	Germany	Latin word "rubidius" meaning "dark red"
1861	Thallium	Tl	81	Sir William Crookes	England	Greek word "thallos" meaning "green twig"
1863	Indium	In	49	F. Reich, H. Richter	Germany	Indigo band in its atomic spectrum
1875	Gallium	Ga	31	P.E.Lecoq de Boisbaudran	France	Latin word "Gallia" meaning "France"
1878	Holmium	Ho	67	J.L. Soret	Switzerland	Greek word "Holmia" meaning "Sweden"
1878	Ytterbium	Yb	70	Jean de Marignac	Switzerland	"Ytterby", a village in Sweden
1879	Scandium	Sc	21	Lars Nilson	Sweden	Latin word "Scandia" meaning "Scandinavia"
1879	Samarium	Sm	62	P.E.Lecoq de Boisbaudran	France	From a mineral "Samarskite"
1879	Thulium	Tm	69	Per Theodor Cleve	Sweden	"Thule", an ancient name for Scandinavia
1880	Gadolinium	Gd	64	Jean de Marignac	Switzerland	"Gadolin", a Finnish chemist
1885	Praseodymium	Pr	59	C.F. Aver von Welsbach	Austria	Greek words "prasios didymos" mean "green twin"
1886	Fluorine	F	9	Henri Moissan	France	Latin word "fluere" meaning "to flow"
1886	Germanium	Ge	32	Clemens Winkler	Germany	Latin word "Germania" meaning "Germany"
1886	Dysprosium	Dy	66	P.E.Lecoq de Boisbaudran	France	Greek word "dysprositos" meaning "hard to obtain"
1894	Argon	Ar	18	W.Ramsey, Baron Rayleigh	Scotland	Greek word "argos" meaning "inactive"
1895	Helium	He	2	W.Ramsey,N.Langet...	Scotland/Sweden	Greek word "helios" meaning "sun"
1898	Neon	Ne	10	W.Ramsey, M.W. Travers	England	Greek word "neon" meaning "new"
1898	Krypton	Kr	36	W.Ramsey, M.W. Travers	Great Britain	Greek word "kryptos" meaning "hidden"
1898	Xenon	Xe	54	W.Ramsay,M. W. Travers	Great Britain	Greek word "xenos" meaning "stranger"
1898	Polonium	Po	84	Pierre and Marie Curie	France	From "Poland"
1898	Radon	Rn	86	Fredrich Ernst Dorn	Germany	From element radium
1898	Radium	Ra	88	Pierre and Marie Curie	France	Latin word "radius" meaning "ray"
1899	Actinium	Ac	89	André Debierne	France	Greek word "aktinos" meaning "ray"
1901	Europium	Eu	63	Eugene Demarcay	France	From "Europe"
1907	Lutetium	Lu	71	Georges Urbain	France	Greek word "Lutetia" meaning "Paris"
1917	Protactinium	Pa	91	F.Soddy,J.Cranston,O.Hahn...	England/France	Greek word "protos" meaning "first"
1923	Hafnium	Hf	72	Dirk Coster,G.von Hevesy	Denmark	Latin name "Hafnia" meaning "Copenhagen"
1925	Neodymium	Nd	60	C.F. Aver von Welsbach	Austria	Wwords "neos didymos" mean "new twin" in Greek
1925	Rhenium	Re	75	W. Noddack,I.Tacke,O.Berg	Germany	Greek word "Rhenus" meaning river "Rhine"
1937	Technetium	Tc	43	Carlo Perrier,Emillo Segre	Italy	Greek word "technikos" meaning "artificial"
1939	Francium	Fr	87	Marguerite Derey	France	From "France"
1940	Astatine	At	85	D.R.Corson,K.R.MacKenzie...	United States	Greek word "astatos" meaning "unstable"
1940	Neptunium	Np	93	E.M.McMillan,P.H.Abelson	United States	From "the planet Neptune"
1940	Plutonium	Pu	94	G.T.Seaborg,J.W.Kennedy...	United States	From "the planet Pluto"
1944	Curium	Cm	96	G.T.Seaborg, R.A.James...	United States	From Pierre and Marie "Curie"
1945	Promethium	Pm	61	J.A.Marinsky,L.E.Glendenin...	United States	From "Prometheus" in Greek mythology
1945	Americium	Am	95	G.T.Seaborg,R.A. James...	United States	From English word "America"
1949	Berkelium	Bk	97	G.T.Seaborg,S.G.Tompson...	United States	From "Berkeley", a city in California,
1950	Californium	Cf	98	G.T.Seaborg,S.G.Tompson...	United States	From "California", USA
1952	Einsteinium	Es	99	Workers at Los Alamos, USA	United States	From "Albert Einstein"
1953	Fermium	Fm	100	Workers at Los Alamos, USA	United States	From "Enrico Fermi"
1955	Mendelevium	Md	101	G.T.Seaborg,S.G.Tompson...	United States	From "Dimitri Mendeleev"
1957	Nobelium	No	102	A.Ghiorso,T.Sikkeland...	Sweden	From Swedish chemist "Alfred Nobel"
1961	Lawrencium	Lr	103	A.Ghiorso,T.Sikkeland...	United States	From "O. Lawrence"
1964	Rutherfordium	Rf	104	Workers at Dubna,Berkeley,USA	Berkeley California	From chemist "Rutherford"
1967	Dubnium	Db	105	Workers at Dubna,Berkeley,USA	Russia/United States	From place "Dubna"
1974	Seaborgium	Sg	106	Albert Ghiorso and others	Russia/United States	From chemist "Glenn T.Seaborg"
1981	Bohrium	Bh	107	P.Armbruster,G.Münzenber	Darmstadt,Germany	From physicist "Niels Bohr"
1982	Meitnerium	Mt	109	P.Armbruster,G.Munzenber	Darmstadt,Germany	From "Meitner", the Austrian physicist
1984	Hassium	Hs	108	P.Armbruster,G.Munzenber	Darmstadt,Germany	From German state "Hess"
1994	Darmstadtium	Ds	110	S.Hofmann,V.Ninov,F.Hessberger...	Darmstadt,Germany	From place "Darmstadt" in Germany
1994	Roentgenium	Rg	111	S.Hofmann,V.Ninov,F.Hessberger...	Darmstadt,Germany	From "Wilhelm Conrad Roentgen"
1996	Ununbium	Uub	112	S.Hofmann,V.Ninov,F.Hessberger...	Darmstadt,Germany	Temporary nomenclature
1998	Ununquadium	Uuq	114	Workers of Nuclear Inst.at Russia	Dubna, Russia	Temporary nomenclature
2000	Ununhexium	Uuh	116	Y.T.Oganessian, V.K.Utyonkov...	Dubna, Russia	Temporary nomenclature

Appendix D

I PAC : 1

American : IA

PERIODIC

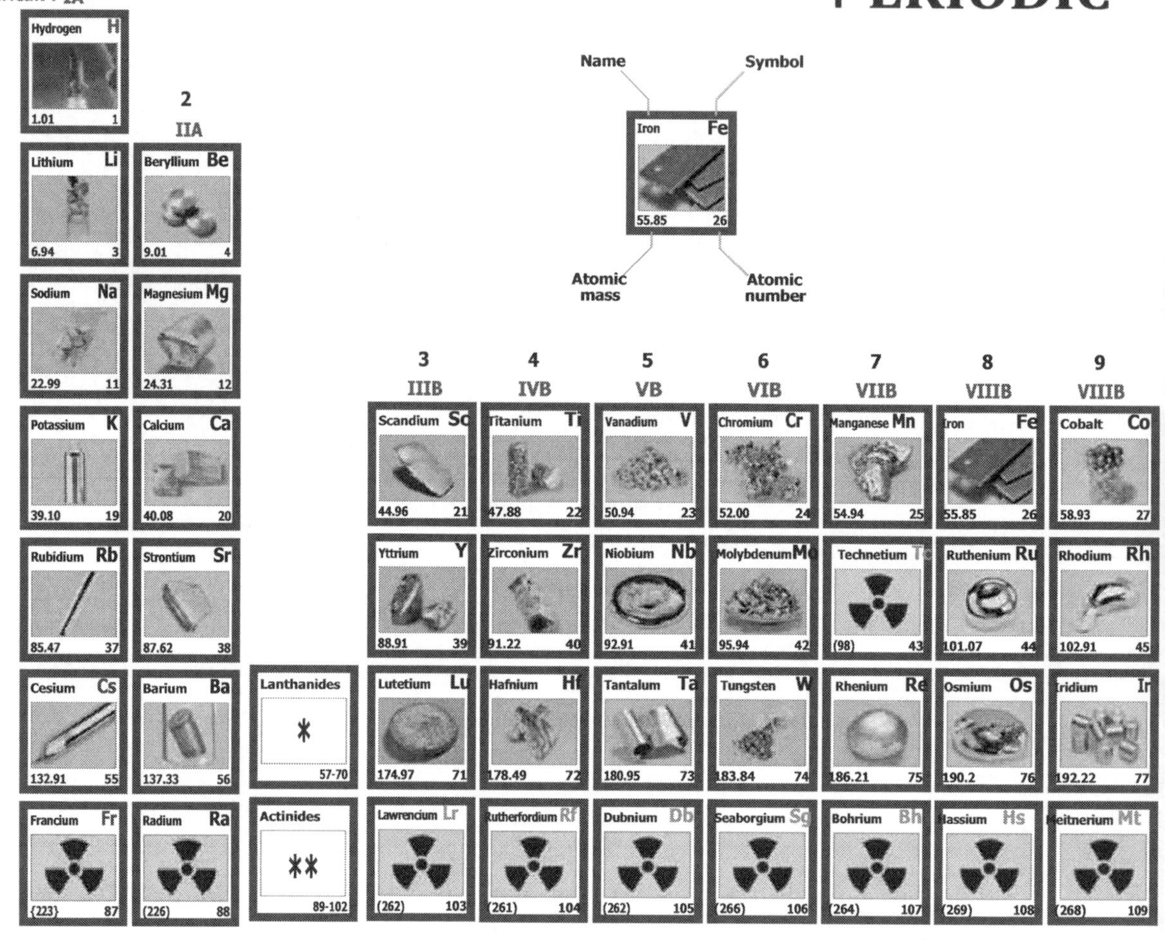

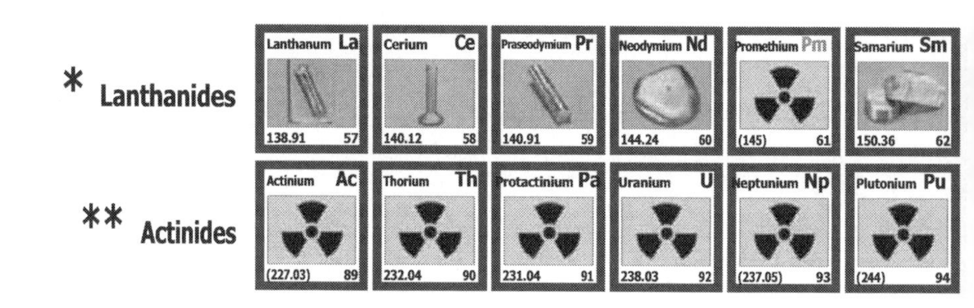

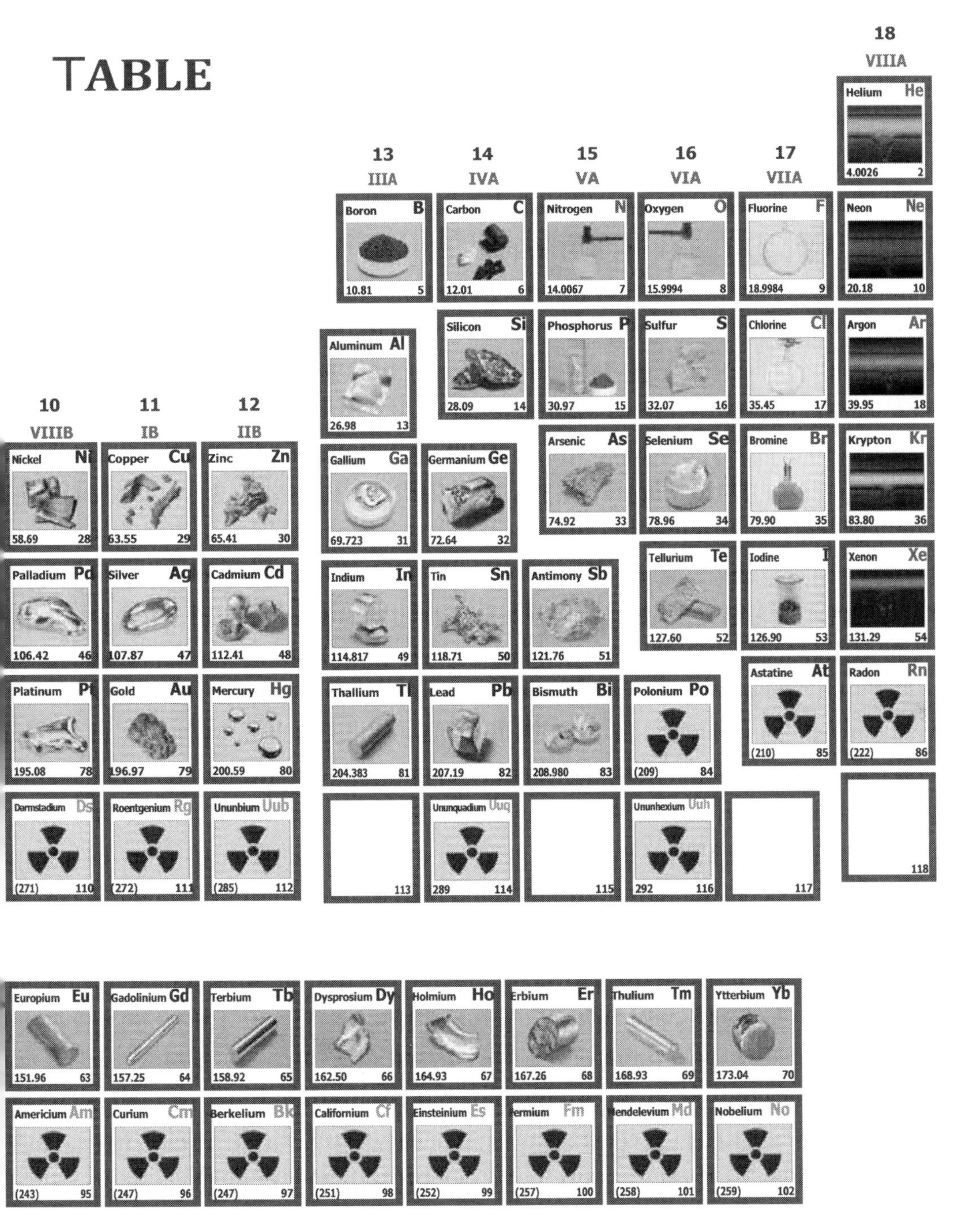
TABLE
18
VIIIA
Helium He 4.0026 2
13 IIIA
14 IVA
15 VA
16 VIA
17 VIIA
Boron B 10.81 5
Carbon C 12.01 6
Nitrogen N 14.0067 7
Oxygen O 15.9994 8
Fluorine F 18.9984 9
Neon Ne 20.18 10
Aluminum Al 26.98 13
Silicon Si 28.09 14
Phosphorus P 30.97 15
Sulfur S 32.07 16
Chlorine Cl 35.45 17
Argon Ar 39.95 18
10 VIIIB
11 IB
12 IIB
Nickel Ni 58.69 28
Copper Cu 63.55 29
Zinc Zn 65.41 30
Gallium Ga 69.723 31
Germanium Ge 72.64 32
Arsenic As 74.92 33
Selenium Se 78.96 34
Bromine Br 79.90 35
Krypton Kr 83.80 36
Palladium Pd 106.42 46
Silver Ag 107.87 47
Cadmium Cd 112.41 48
Indium In 114.817 49
Tin Sn 118.71 50
Antimony Sb 121.76 51
Tellurium Te 127.60 52
Iodine I 126.90 53
Xenon Xe 131.29 54
Platinum Pt 195.08 78
Gold Au 196.97 79
Mercury Hg 200.59 80
Thallium Tl 204.383 81
Lead Pb 207.19 82
Bismuth Bi 208.980 83
Polonium Po (209) 84
Astatine At (210) 85
Radon Rn (222) 86
Darmstadium Ds (271) 110
Roentgenium Rg (272) 111
Ununbium Uub (285) 112
113
Ununquadium Uuq 289 114
115
Ununhexium Uuh 292 116
117
118
Europium Eu 151.96 63
Gadolinium Gd 157.25 64
Terbium Tb 158.92 65
Dysprosium Dy 162.50 66
Holmium Ho 164.93 67
Erbium Er 167.26 68
Thulium Tm 168.93 69
Ytterbium Yb 173.04 70
Americium Am (243) 95
Curium Cm (247) 96
Berkelium Bk (247) 97
Californium Cf (251) 98
Einsteinium Es (252) 99
Fermium Fm (257) 100
Mendelevium Md (258) 101
Nobelium No (259) 102

Appendix E

MENDELEEV'S PERIODIC TABLE

Groups / Periods	A I B	A II B	A III B	A IV B	A V B	A VI B	A VII B	A VIII B		
1	(H)						1 H	2 He		
2	3 Li	4 Be	5 B	6 C	7 N	8 O	9 F	10 Ne		
3	11 Na	12 Mg	13 Al	14 Si	15 P	16 S	17 Cl	18 Ar		
4	19 K	20 Ca	21 Sc	22 Ti	23 V	24 Cr	25 Mn	26 Fe	27 Co	28 Ni
	29 Cu	30 Zn	31 Ga	32 Ge	33 As	34 Se	35 Br	36 Kr		
5	37 Rb	38 Sr	39 Y	40 Zr	41 Nb	42 Mo	43 Tc	44 Ru	45 Rh	46 Pd
	47 Ag	48 Cd	49 In	50 Sn	51 Sb	52 Te	53 I	54 Xe		
6	55 Cs	56 Ba	57 La *	72 Hf	73 Ta	74 W	75 Re	76 Os	77 Ir	78 Pt
	79 Au	80 Hg	81 Tl	82 Pb	83 Bi	84 Po	85 At	86 Rn		
7	87 Fr	88 Ra	89 Ac **	104 Rf	105 Db	106 Sg	107 Bh	108 Hs	109 Mt	110 Ds

He — symbol; 2 — atomic number

s element
p element
d element
f element

Lanthanides *	58 Ce	59 Pr	60 Nd	61 Pm	62 Sm	63 Eu	64 Gd	65 Tb	66 Dy	67 Ho	68 Er	69 Tm	70 Yb	71 Lu
Actinides **	90 Th	91 Pa	92	93 Np	94 Pu	95 Am	96 Cm	97 Bk	98 Cf	99 Es	100 Fm	101 Md	102 No	103 Lr

GLOSSARY

Actinides : The series of elements coming after actinium and before element 104 in the periodic table.

Activation energy : The minimum amount of energy required to initiate a reaction.

Alkali metal(s) : The group 1A elements of the periodic table.

Alkaline earth metal(s) : The group 2A elements of the periodic table.

Allotrope(s) : Different forms of the same element that have different molecular structures but often exist in the same physical state.

Alpha particle : Positively charged particle emitted at high speeds from certain radioactive substances (helium nuclei).

Artifical element : Not found in nature (produced in the laboratory).

Atomic mass : The mass of an atom expressed in atomic mass units.

Aufbau Principle : In building up the electron configuration of an atom or a molecule in its ground state, the electrons are placed in the orbitals in order of increasing energy.

Binding energy : The energy required to separate a nucleus into its components, protons and neutrons

Cathode ray tube : A cathode ray tube, usually made of glass, from which most of the air has been removed and inside which a beam of electrons is produced.

Coulomb's Law : The magnitude of the electric force that a particle exerts on another particle is directly proportional to the product of their charges and inversely proportional to the square of the distance between them.

Covalent bond : The bond in which two electrons are shared by two atoms.

Dalton's Atomic Model : Elements are composed of extremely small particles called 'atoms'. All atoms of the same element are identical. Atoms are filled spheres which cannot be further divided.

Dalton's Law : In a mixture of gases, the total pressure of the mixture is equal to the sum of the pressures that each gas would exert by itself in the same volume.

d - block : The portion of the periodic table in which the process of filling electron orbitals involves d orbitals.

Density : The mass of a unit volume of a substance.

Electromagnetic Spectrum : The spectrum of all radiation resulting from fluctuations of electric currents and vibrations of charged particles. Electromagnetic radiation travels at the speed of light and includes visible light, X-rays, ultraviolet light, infrared light, radio waves, etc.

Electron : Negatively charged subatomic particle.

Electron capture : A nuclear decay process in which an electron outside the nucleus is captured and a proton is converted to a neutron inside the nucleus.

Electron affinity : The energy change when an electron is accepted by an atom (or an ion) in the gaseous state.

Electron configuration : A shorthand notation describing the distribution of electrons among the subshells of an atom.

Electronegativity : A measure of the tendency of an atom in a stable molecule to attract electrons within bonds.

Energy level (shell) : A shell which is one of the successive layers of electrons around an atom.

Excited state : A state of an atom, molecule, etc, when it has absorbed energy and become excited compared to the ground state.

f - block : The portion of the periodic table in which the process of filling electron orbitals involves f orbitals. These are the lanthanide and actinide elements.

Ground state : The state in which all the electrons in an atom are in the lowest energy levels available.

Halogens : The group 7A elements in the periodic table.

Hund's rule : Whenever orbitals of equal energy are available, electrons are assigned to these orbitals singly before any pairing of electrons occurs.

Ionic bond : The bond formed between a metal and a nonmetal as a result of electron transfer.

Ionization energy : The energy required to remove an electron from an atom or ion in a gaseous state.

Isoelectronic : Molecules or ions which have the same number of electrons, e.g. CO and N_2.

Lanthanides : The series of elements coming after lanthanum and before hafnium in the periodic table.

Law of conservation of mass : The total mass of substance or substances that exist before a chemical reaction is equal to the total mass of the substance or substances that exist after the reaction, i.e., matter can not be neither created nor destroyed.

Law of definite proportion : The propotion by mass of the elements in a given compound is always the same, regardless of its source or manner of preparation.

Law of multiple proportion : Deals with the proportions in which two elements form two or more compounds for each compound the ratio of the masses of one element to a fixed mass of the second element is formulated. These ratios are in the ratio of small whole numbers

Lewis structure : Writing of formulas by representing valence electrons by dots.

Liquid pressure : The force exerted by a liquid on per unit area; determined by the formula of $p = h \cdot d$

Melting point : The temperature at which solid and liquid phases coexist in equilibrium.

Metal : An element that forms positive ions in chemical reactions.

Metalloid (semimetal): An element with both metallic and nonmetallic properties.

Neutron : Subatomic particle which has not any charge.

Neutral atom : An atom which has the same number of protons and electrons.

Noble gases : The elements helium, neon, argon, krypton, xenon and radon. Family of inactive gases found in group 8A of the periodic table.

Nonmetal : An element (in the upper right corner of the periodic table) that does not show metallic properties and generally, form negative ions in chemical reactions.

Normality : The number of equivalents of solute per liter of solution.

Nucleus (pl.nuclei) : The very small, very dense, positively charged center of an atom containing protons and neutrons as well as other subatomic particles.

Octet : A stable outer shell of eight electrons, arranged as four orbital pairs.

Oxidation : That part of an oxidation reduction reaction characterized by electron loss or by an algebraic increase in oxidation numbers.

Oxidation state number : Charge of an atom in a compound and charge of element.

Pauli's Principle : When two electrons occupy the same orbital, they must have opposite spins.

p - block : The portion of the periodic table in which the filling of electron orbitals involves p orbitals.

Period : A horizontal sequence of the periodic table. Following the first period of H and He, the others range from an alkali metal on the left to a noble gas on the right.

Periodic table : A table summarizing a great deal of information about the elements, arranging them by atomic number.

Photon : A "particle" of light. The energy of a beam of light is concentrated into these photons.

Proton : Positively charged subatomic particle.

Positron: A nuclear particle having the same mass as an electron but a positive charge.

Quark : Fundamental subatomic particle.

Rutherford's atomic model : Just like the solar system, an atom has a nucleus, consisting of positively charged particles, around which electrons orbit.

s - block : The portion of the periodic table in which the filling of electron orbitals involves s orbitals of the outermost shell.

Temperature : A physical property that determines the direction of heat flow in an object on contact with another object. It is a measure of the average kinetic energy of the molecules.

Thomson's atomic theory : Electrons randomly occupy the surface of an atom like the grapes on the surface of a cake.

Transition elements : Some elements that lie in rows 4–7 of the periodic table, comprising scandium through zinc, yttrium through cadmium, and lanthanum through mercury.

Transuranium elements : The elements following uranium (atomic number 92) in the periodic table.

Valence electrons : The electrons in the outermost shell (valence shell) of an atom.

ANSWERS

SUPPLEMENTARY QUESTIONS

THE HISTORY OF THE ATOM

28. a. s orbital b. p orbitals
    c. d orbitals d. f orbitals
29. a. 6 electrons from p orbitals
    b. 6 electrons from p and 10 electrons from d orbitals, totally 16 electrons
    c. 2 electrons from s and 14 electrons from f orbitals, totally 16 electrons
30. Totally 2 electrons (one of the f orbitals)
31. 9
32. a. ml = 0 b. ml = −1,0,+1
    c. l = 1,2 d. n = 2,3,4...
33. a. n = 3 l = 0
    b. n = 4 l = 1
    c. n = 5 l = 3
    d. n = 3 l = 2
34. a. 1 b. 5 c. 3 d. 7 e. 5
36. a. $_{20}Ca^{2+}$: [Ar] b. $_{30}Zn^{2+}$: $[Ar]4s^2 3d^8$
    c. $_{25}Mn^{2+}$: $[Ar]4s^2 3d^3$ d. $_{26}Fe^{3+}$: $[Ar]4s^2 3d^3$
    e. $_{17}Cl^-$: [Ar] f. $_{34}Se^{2-}$: [Kr]

THE PERIODIC TABLE

3. According to atomic masses of elements
9. The first period
10. 2, 8, 8, 18 and 18
11. Hydrogen (H) and helium(He)
13. Li, Na, K, Rb, Cs and Fr
15. Be, Mg, Sr, Ba and Ra
16. Helium, neon, argon, krypton, xenon and radon
17. 1A, 2A, 7A and 8A
19. $_5B$: $1s^2 2s^2 2p^1$
    $_{15}P$: $1s^2 2s^2 2p^6 3s^2 3p^3$
    $_{30}Z$: $1s^2 2s^2 2p^6 3s^2 3p^6 4s^2 3d^{10}$
    $_{31}Ga$: $1s^2 2s^2 2p^6 3s^2 3p^6 4s^2 3d^{10} 4s^1$
20. $[_{36}Kr]\ 4d^{10} 5s^1$
21. a) 1st period, group 8A
    b) 2nd period, group 7A
    c) 4th period, group 5B
    d) 3rd period, group 3A
    e) 4th period, group 2B
    f) 4th period, group 7A
22. 11
23. $_3Li$, $_7N$
24. Al : 1, Si : 2, P : 3, Mn : 5
25. 31
26. 8B
27. 8, 12 and 15
28. a) 35 protons, 45 neutrons
    b) 4
    c) 5
    d) p orbitals
29. Rb > Zn >V > S
31. b) $_{16}S^{2-}$, $_{20}Ca^{2+}$, $_{15}P^{3-}$, $_{19}K^{1+}$ and $_{25}Mn^{7+}$
32. 3rd period, group 6A
33. 4
34. Mg : 2, O : 6, Si : 4 and Zn : 2
35. Tin (Sn)
36. a) Noble gas b) Metal
    c) Metal d) Metal
    e) Metalloid f) Nonmetal
    g) Metal h) Metal
37. K > Na > Mg > C
38. Na
39. Rn
41. Ca
42. $_7N > {}_8O > {}_4Be > {}_5B$

43. ${}_{10}Ne > {}_{9}F > {}_{11}Na$
44. 109.6 kJ/mol
45. ${}_{11}Na > {}_{10}Ne > {}_{17}Cl$
46. Elements of group 1A
47. Mg^{2+}
53. F > Cl > Si > Mg
54. III = IV > V > II > I
55. CO neutral; CO_2,SO_2,SO_3 acidic; CaO basic; SiO_2 amphoteric.
56. K
57. Cl

RADIOACTIVITY

4. Z is radioactive
14. a. β^- b. α c. β^+ d. n
15. a. A = 230 b. Z = 11 c. A = 14, Z = 6 d. A = 209, Z = 83
16. Atomic mass number is decreased by 4. Atomic number does not change.
17. 138
18. 138
19. 90 ; 234
20. 92
21. 5A
22. 3
23. X and T are isotope
24. Neutron
25. 9
26. 33
27. XY_3
28. 11 ; 13
30. 30 hours
31. 8 hours
32. 12 days
33. 64 g
34. 3.5 g
35. 24 g
36. 48 g
37. 30 g
38. 12.5 g
39. 20 minutes
40. 80 g X and 20 g Y
46. Scorpion
47. β^- rays
48. Americium - 241
49. Gamma rays (ψ)
50. carbon -14
51. ψ rays kill cancer cells
52. X - rays

MULTIPLE CHOICE

THE HISTORY OF THE ATOM

1. B
2. D
3. C
4. D
5. C
6. B
7. C
8. A
9. B
10. D
11. B
12. D
13. C
14. D
15. D
16. C
17. D
18. A
19. A
20. A
21. E
22. D
23. C

THE PERIODIC TABLE

1. D
2. B
3. E
4. D
5. C
6. B
7. D
8. B
9. D
10. A
11. C
12. D
13. C
14. C
15. A
16. A
17. C
18. D
19. E
20. E
21. E
22. D

RADIOACTIVITY

1. D
2. C
3. C
4. C
5. D
6. A
7. E
8. D
9. C
10. B
11. E
12. C
13. E
14. D
15. B
16. E
17. C
18. D
19. B
20. C
21. C
22. B
23. E
24. E
25. B

PUZZLES

THE HISTORY OF THE ATOM

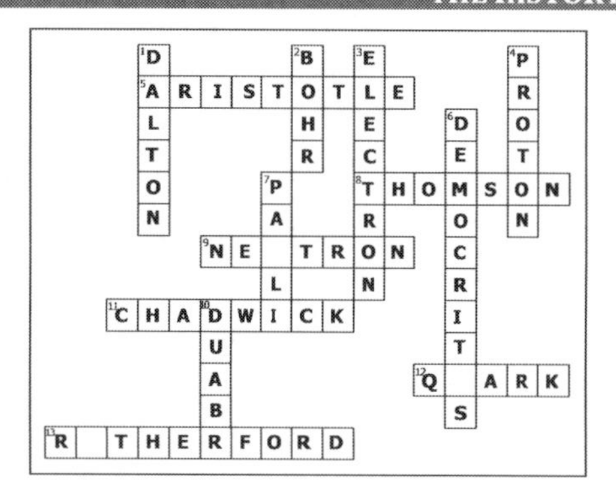

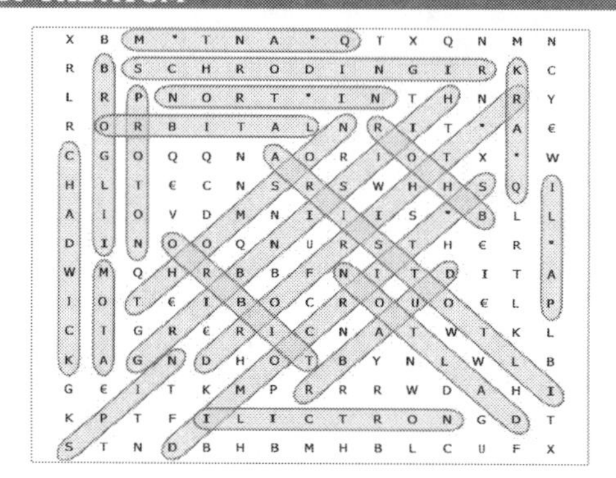

THE PERIODIC TABLE

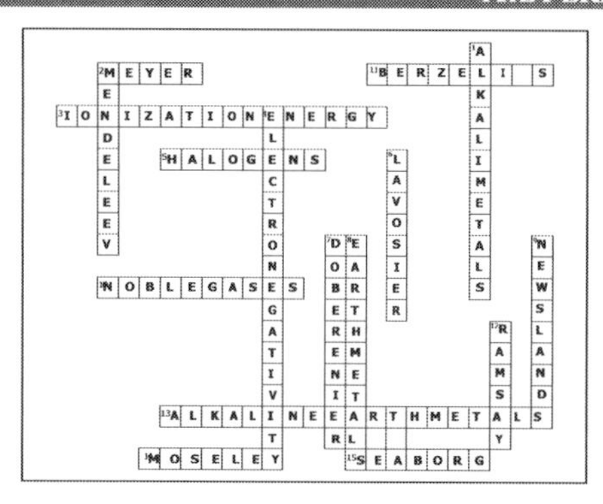

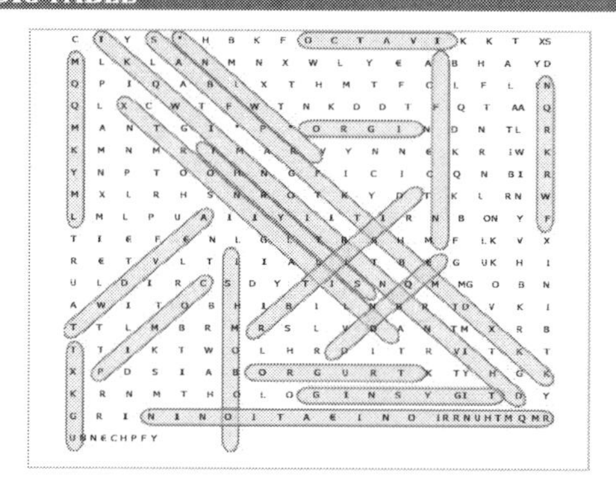

RADIOACTIVITY

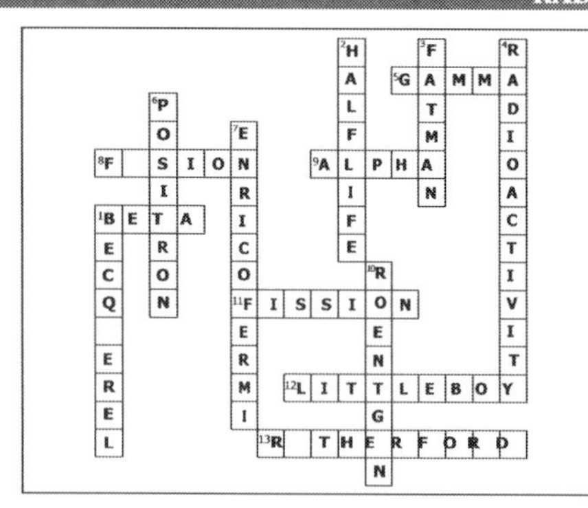

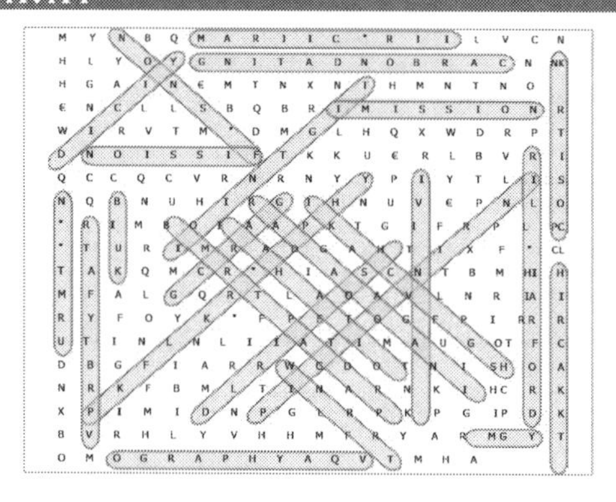

INDEX

REFERENCES

Aybers, N. and others. **The Economic Comparison of Nuclear Reactors of 350 and 500 MW(e) for Developing Countries.** İstanbul Teknik Üniversitesi matbaasý, İstanbul : 1994

Bekaroðlu, Prof. Dr. Özer. **Genel Kimya Teori ve Problemleri.** Kipaþ, İstanbul : 1986

Berkem, Prof. Dr. Ali Rýza. **Çekirdek Kimyasý ve Radyokimya.** İstanbul Üniversitesi Yayýnlarý, İstanbul : 1992

Chang, Raymond. **Chemistry international edition.** Mc Graw – Hill Inc. USA : 1984

Chaudhry, Dr. Rehman and others. **Chemistry for College.** Punjab Textbook Board, Lahor : 1996

Darlington, C. Le Roy and others. **The Chemical World, Activities and Explorations.** Mifflin Company, USA : 1987.

Dorin, Henry and others. **Chemistry, the study of matter.** Prentice Hall Inc., New Jersey : 1992

Frank, D.V.; Little J.G.; Miller, S. **Chemical Interactions.** Prentice Hall Science Explorer, Pearson Education Inc., New Jersey : 2005

Gallagher, RoseMarie; Ingram, P. **Modular Science, Chemistry.** Oxford University Press : 2001

Gillespie, Ronald J. and others. **Chemistry second edition.** Allyn and Bacon Inc. Massachusetts : 1989

Haire, Marian; Kennedy, E; Lofts, G; Evergreen, M.J... **Core Science.** John Wiley & Sons Australia Ltd. Publication : 1999

Hein, Morris & Best, Leo **College Chemistry.** Brooks and Cole Publishing Company, Monterey, California : 1986

Holltzchaw, H. F. and Robinson, W. R. **Study Guide For General Chemistry and College Chemistry.** D.C. Heath and company. Lexington, Massachusetts : 1988

Masterton, William L. and Hurley, Cecile. **Chemisty, Principles.** Saunders College publishing, Philadelphia : 1989

Mortimer, Charles E. **Chemistry.** Wadsworth Publishing Company, England : 1987

Oxtoby, David W. and others. **Chemistry, Science of change.** Saunders College Publishing,USA : 1991

Öztürk, Gürkan. "Mikro dünyanýn yeni pencereleri" **Bilim ve Teknik Dergisi.** Kasým, 1991. sayý : 288, ss. 38 – 43

Petrucci, Ralph H. **General Chemistry, Principles and Modern Applications.** Mac Millan Publishing Company, New York : 1989

Prescott, C.N. **CHEMISTRY, A Course for "0" Level.** Federal Publications. Singapore : 2000

Rudzitis, Q.Y and Feldman, F.G. **Chemistry (10–11)**. Prosveshenie, Moscow : 1996

Rudzitis,Q.Y and Feldman,F.G. **Chemistry (8-9)** Maarif, Baku : 1996

Ryan, L. **Chemistry for You.** Stanley Thornes (Publishers) Ltd. England : 1996

Sevenair, J.P.; Burkett A.R. **Introductory Chemistry.** WM.C. Brown Publishers : 1997

Watkins, Patricia and others. **General Science.** Harcourt Brace Jovanich Publishers, Orlando : 1983

Whitten, Kenneth W. and others. **General Chemistry.** Saunders College Publishing, New York : 1992

Printed in Great Britain
by Amazon